A. Tschirch, Friedrich Flückiger

The Principles of Pharmacognosy

An Introduction to the Study of the Crude Substances of the Vegetable Kingdom

A. Tschirch, Friedrich Flückiger

The Principles of Pharmacognosy
An Introduction to the Study of the Crude Substances of the Vegetable Kingdom

ISBN/EAN: 9783337171063

Printed in Europe, USA, Canada, Australia, Japan

Cover: Foto ©berggeist007 / pixelio.de

More available books at **www.hansebooks.com**

THE

PRINCIPLES

OF

PHARMACOGNOSY

AN INTRODUCTION TO
THE STUDY OF THE CRUDE SUBSTANCES
OF THE VEGETABLE KINGDOM

BY

FRIEDRICH A. FLÜCKIGER, Ph.D., M.D.,

PROFESSOR IN THE UNIVERSITY OF STRASSBURG

AND

ALEXANDER TSCHIRCH, Ph.D.,

LECTURER ON BOTANY AND PHARMACOGNOSY IN THE UNIVERSITY OF BERLIN

WITH ONE HUNDRED AND EIGHTY-SIX ILLUSTRATIONS IN THE TEXT

TRANSLATED FROM THE
SECOND AND COMPLETELY REVISED GERMAN EDITION

BY

FREDERICK B. POWER, Ph.D.,

PROFESSOR OF MATERIA MEDICA AND PHARMACY IN THE UNIVERSITY OF WISCONSIN

NEW YORK
WILLIAM WOOD & COMPANY
1887

THE AUTHORS' PREFACE.

In the present work, the "Principles of Pharmaceutical Materia Medica" (*Grundlagen der Pharmaceutischen Waarenkunde*), by F. A. Flückiger, published in 1873, appear in a revised form, and materially extended. In this second edition, the above-named author has undertaken chiefly the preparation of the first part, and the newly associated author (A. Tschirch), the second part, comprising morphology and anatomy, so that really a new book has been produced. Although the original plan has been essentially retained, the arrangement of the material within the more extended space has, nevertheless, necessitated numerous changes.

It seemed to us appropriate in this revision to also take into consideration those crude substances of the vegetable kingdom possessing an organic structure which receive technical application. On the other hand, zoology has now been but incidentally considered, for the reason that pharmacognosy has to treat of only a very limited number of animal substances.

With regard to the classification of tissues, we have thought it proper to follow Haberlandt's "Physiological Plant Anatomy" (*Physiologische Pflanzenanatomie*), Leipzig, 1884, which, besides the anatomical consideration of tissues, also elucidates their physiological functions. In our capacity as teachers, we have acquired the experience that an anatomical description becomes of much more interest to the pupil by a reference to physiological relations. Although the system

of physiological plant anatomy in its details is still in need of completion and general recognition, the outlines of the same appear to us, nevertheless, to be sufficiently well defined to be able to serve, in the manner intimated, as a foundation.

In one point, however, we have departed from Haberlandt's classification. The cell, its contents, and its membrane have been treated by us in a more complete manner, and placed before the tissues. This deviation recommends itself for practical and didactic reasons; for the beginner must be instructed regarding the cell and its contents before learning anything of the tissues. The substances contained in the cell also possess too great an interest for the pharmacognosist to be inserted in the text simply in a secondary manner.

A second change, as already intimated, is that we have likewise made a place for technico-microscopical investigation, of which every one will approve who takes into consideration the extent to which the apothecary of our day is called upon as an expert. In order to aid in furthering this service of pharmacy, which is of general interest, we have, for example, treated more thoroughly of starch and the textile fibres.

In the morphological portion, which has experienced a complete transformation, we have endeavored to give a brief sketch of the most important phenomena, whereby the technical expressions in present use have received explanation, especially those which occur in the deservedly widely distributed *Syllabus* of Eichler. Here, as in the anatomical portion, we have drawn the narrowest boundaries, since the present work does not pretend to be a complete text-book or manual. Nevertheless, some few sections which are of importance to pharmacy have received relatively somewhat greater development and more precise adaptation, as for example, that relating to the receptacles for secretions. A chapter on the galls has also been newly inserted.

In the selection of the woodcuts, we have chiefly considered the drugs and crude substances of technical application. The same applies to the selection of examples from anatomy and morphology. The one hundred and four illustrations of the first edition have, for the most part, been again introduced, although increased by thirty-seven illustrations sketched by the newly associated author, as also by a number borrowed from other works, the names of which, as a rule, have been mentioned. With some of the illustrations which were placed at our disposal by the publishing house of Springer, it has not each time been stated whether these were derived from the works of Hager, Hartig, or Möller. Very often we were also obliged to content ourselves by referring to still other illustrations, for example, to the plates of the handsome anatomical atlas by Berg.

It has likewise been our endeavor, through abundant citations of the literature, to be of service to those members of the profession who may desire more complete information. Although not every one will be in a position to refer to the sources of information cited by us, we nevertheless wished, on the one hand, to show that we have endeavored to utilize the best and most recent acquisitions upon the wide field of the auxiliary sciences, and, on the other, we hope thereby to afford an impulse in many directions. The latter purpose was particularly kept in view with regard to the section devoted to the history of drugs.　　　　　　　　　　　　THE AUTHORS.

STRASSBURG AND BERLIN, May, 1885.

TRANSLATOR'S PREFACE.

THE generally acknowledged importance of the study of pharmacognosy as a branch of useful knowledge, and the constantly increasing recognition of its extended practical, as well as scientific applications, will doubtless afford to all who are conversant with the subject a sufficient vindication for the production of an English version of a work which bears, as its highest commendation, the names and impress of its distinguished authors.

The work here presented will doubtless at once indicate the aim and the scope of the science of pharmacognosy, and clearly demonstrate its intimate connection on every hand with chemical, botanical, and microscopical science, as also the impulse afforded for the further investigation of points of historic interest and a more extended knowledge of the geographical or climatic and commercial relations of vegetable products in their varied applications, either as medicinal remedies, as food, or in the arts.

However important the consideration of the physiological action and therapeutic uses of drugs may be from the standpoint of medical science, it is evident that this alone should not suffice for the professional pharmacist or for pharmaceutical students, who should be instructed in the science of pharmacognosy in its broadest sense. It is thus to be hoped that the principles outlined in the work in question may serve to broaden the exposition of the organic materia medica in the directions intimated, and to secure for the same a still wider recognition and better appreciation of its usefulness.

In the preparation of this edition, the translator has been kindly favored by Professor Flückiger with a few additional notes, which have been suggested by the advances in literature or in the related sciences during the short period which has elapsed since the publication of the German edition, and a list of the illustrations occurring in the work has also been appended.

The translator may finally be permitted to state that he has endeavored to preserve in this edition not only the form, but also the attractiveness of the original work, and while attempting to follow the original as closely as was consistent or possible, has spared no pains to maintain accuracy of diction and the correct rendition of many newly adopted technical terms.

That the work may accomplish the mission designed by its authors, receive to some degree the appreciation which it merits, and be made available to a larger circle of readers, was the highest purpose and the chief desire attending the labor which the translator has been permitted to bestow.

UNIVERSITY OF WISCONSIN, MADISON, January, 1887.

SYNOPSIS OF CONTENTS.

LIST OF ILLUSTRATIONS.

xvi LIST OF ILLUSTRATIONS.

THE PRINCIPLES

OF

PHARMACOGNOSY.

THE MISSION OF PHARMACOGNOSY.

THE substances which are employed medicinally for their remedial action are either the products of human skill or they belong directly to the two organic natural kingdoms. Among the medicinal substances formed through chemical operations we meet, indeed, with such as are produced only by certain definitely conducted chemical or physical processes, as, for example, the acids, iodine, bromine, chloral, phenol, glycerin, alcohol, and the alkaloids, be it that chemical industry starts from materials belonging to the inorganic kingdom, or that its first point of departure, as in the two last-named examples, falls within the circle of organic nature. In some rare cases only (mineral waters, for instance) is it sufficient to merely choose a suitable form for that which nature offers ready for use; more frequently chemical skill is directed toward the isolation of active principles from plants (exceptionally also from animals or animal substances, or from the mineral kingdom), and to the separation of these principles from associated substances, or, in other words, to their purification. In all these cases the object is to place at the disposal of medical science bodies which

1

are sharply characterized in a chemical sense, or, briefly stated, chemical units; for only such substances as are at all times accessible and complete in their identity can afford a sure foundation for scientific medicine and pharmacy. In this direction lies the aim of the future.

Medicinal agents of this kind are outside of the sphere of pharmacognosy. By general concurrence in pharmaceutical circles, there are assigned to it those substances which are directly furnished by nature, or at least such as have not actually been submitted to chemical processes. Since the few crude medicinal substances from the mineral kingdom which were employed in former times have long ago lost their significance, the scientific knowledge which pharmacognosy has to offer is confined to organic nature, or virtually only to the vegetable kingdom; for even among the animals, and the parts and products of animals, only *castor*, *musk*, and *cantharides* represent at the present time important elements as medicinal agents. In the *cantharides* cantharidin alone is of importance, which now stands at the disposal of medical science in a pure form. Pharmaceutical interest in the beetle itself is therefore diminished in a similar manner as it is, at least from this standpoint, in the animals which afford *cod-liver oil, honey, isinglass*, and *milk-sugar*.

The wonderful development of the natural sciences and of medicine, which exerts so great an influence, especially since the second decade of the present century, has liberated pharmacognosy of an enormous burden. In a pharmaceutical work which was first published at Ulm in the year 1641, under the title "Pharmacopœia medico-chymica seu Thesaurus pharmacologicus," by the city physician Johann Christian Schröder, of Frankfurt-on-the-Main, and which in its time was much valued, the author also enumerated the "simplicia" then employed. Among these there were about thirty minerals, and more than one hundred and fifty medicinal substances derived from the animal kingdom or representing entire animals, together with a very large number of roots, herbs, leaves, etc. Such a superfluity of medicinal agents, with all of which it was impossible to become accurately acquainted, characterizes the medicine and pharmacy of the European middle ages [1] and the condition of popular medicine still existing among the Asiatic nations.

The primary object of modern pharmacognosy is to scrutinize all the evidence offered by botany, zoology, and pharmacy regarding the medicinal agents under consideration, to arrange this material in scientific form, and by an appropriate and comprehensive representation to subject it to a closer examination. It is only by this means that pharmacognosy assumes the form of a branch of knowledge which is of equal importance to both pharmacy and medicine. Upon a thorough acquaintance with medicinal substances, and their proper treatment, the practical success of pharmacy is largely dependent, so that a deeper study of the science of pharmacognosy may reasonably be expected of the pharmacist. He will therefore confer honor upon himself and his profession if he advances somewhat farther even than is unavoidably required by his immediate interests. It is, however, scarcely possible to sharply define the boundary between the daily round of duties and scientific studies, and, indeed, this is not only impracticable, but even undesirable.

Pharmacognosy should therefore comprehend, to a certain degree, all that pertains to a monographic knowledge of medicinal substances.[2] The consideration of these substances from the above-outlined standpoint of natural science is to be supplemented by purely historical and geographical references, and such as are connected with the history of civilization, together with commercial relations. All this should be made to assume a rich, animated, and symmetrical form which, in many cases, may become quite attractive. Among this copious material, however, those characters must be prominently pointed out which can lead to a rapid, approximate valuation, whenever it is

[1] Compare "Die Frankfurter Liste" in Archiv der Pharm., 201 (1872), pp. 453–511, and "Das Nördlinger Inventar," *ibid.*, 211 (1877), p. 97.

[2] The aims of modern pharmacognosy have recently been thoroughly explained in the essay: "Ueber die Bedeutung der Pharmacognosie als Wissenschaft, etc.," by A. Tschirch, Pharm. Zeitung, 1881, No. 9.

practicable to do this, without resorting to an actual chemical
analysis. In an accurate knowledge of the nature of medicinal
substances lies, indeed, the best protection against substitution
and adulteration.

The most important property of medicinal substances, however, is their medicinal action, yet this must remain excluded
from pharmacognostical consideration, for it has become the
subject of an independent scientific discipline, viz., pharmacology. From apparent practical reasons it will, indeed, occasionally be found advisable, especially in the case of particularly
interesting substances, to at least allude to their therapeutic
action. That these two domains present many points of contact,
and that pharmacology receives especial support from scientific
pharmacognosy, in fact, that it presupposes a knowledge of the
latter, is clearly evident. In England, France, and other countries, the two expressions, *Pharmacology,* which treats of the
therapeutic action of medicinal substances, and *Pharmacognosy,*
which comprehends a scientific knowledge of the substances
themselves, are occasionally employed in a sense different from
that which has just been indicated. It must be admitted that
the tenor of the two words expresses no sharp distinction.[1]

With regard to the chemical side of pharmacognosy, it is
necessary to restrict it within certain limits. It is doubtless
quite as appropriate that the properly isolated constituents of
drugs should be enumerated and characterized, as that it should
be stated, or at least intimated, where material deficiencies occur
in this direction, which it is desirable to supply. An exhaustive
treatment of the chemical constituents, however, falls within
the domain of chemistry, or pharmaceutical chemistry.

[1] Schmiedeberg, in his "Grundriss der Arzneimittellehre," Leipzig,
1883, says: "The science which treats of medicinal substances (*Arzneimittellehre*) has only to do with such agents as are useful in the curing
of disease. It is, therefore, desirable that all substances which do not
serve as food, and which, through their chemical properties, produce
changes in the living animal organism, should, in order to investigate
these effects, be brought within the borders of a single science, which
may be called *Pharmacology,* or, since it is chiefly supported by experiment, *Experimental Pharmacology.*"

Accordingly, the aim and position of pharmacognosy appear definitely outlined when, beside the questions which it has to answer, the boundaries are also drawn beyond which it should not pass. It is thus to be understood, when on page 3, with a less exact expression, mention was made of a "certain degree" of completeness.

In the fostering of pharmacognosy, botanists and physicians in former times have rendered service, aided, indeed, occasionally by apothecaries.[1] The most brilliant of such services on the part of physicians was brought to a close by the publication in London, 1839 and 1840, of "The Elements of Materia Medica and Therapeutics" by Jonathan Pereira; for, in the mean time, the separate and more thorough treatment of pharmacognosy in the previously explained direction had been taken in hand by pharmacists, and at the earliest period, and with by far the greatest success, by Guibourt, the former teacher of this branch of science at the École de Pharmacie in Paris.[2] Similar, though less prominent results were accomplished in Germany by Johann Bartholomaeus Trommsdorff, of Erfurt, through his "Handbuch der pharmaceutischen Waarenkunde," Gotha, 1822, and particularly, also, by Theodor Wilhelm Christian Martius. In his "Grundriss der Pharmakognosie des Pflanzenreiches," Erlangen, 1822, Martius says that pharmacognosy is to be regarded as "a part of general materia medica, or that science which relates to the examination of the medicinal substances derived from the three kingdoms of nature with a view to ascertain their source and quality, to test them for their purity, and to

[1] Examples in Flückiger's "Pharmakognosie," 2d edition, p. 1,013 (Pires); p. 992 (Morgan); p. 209 (Bansa).

[2] As precursors of Guibourt may be mentioned Nicholas Lémery, author of the "Traité universel des Drogues et Simples," 1697, and Etienne François Geoffroy, whose "Tractatus de Materia Medica" appeared in 1741. These two Parisian apothecaries were, however, more properly recognized as physicians. The "Histoire générale des Drogues," 1694, which is likewise to some extent worthy of consideration, has as its author the druggist Pierre Pomet. The three above-named publications demonstrate how much pharmacognosy was at that time indebted to Paris.

determine substitutions and adulterations." In the year 1825, Martius began to deliver lectures on pharmacognosy at the University of Erlangen, which were the first, as it appears, that were announced under this name. The credit of having aided in securing general approbation for the new expression *pharmacognosy*[1] is due to the writings and lectures of Wiggers (see below).

The question, however, now arises: what is a medicinal substance? To attempt to give a more precise definition is useless, for this term has been undergoing a continuous change in the course of time and in the light of advancing knowledge, not only from the period of antiquity, but also from country to country, and, indeed, from one medical school to another, from one pharmacopœia to another. We are compelled here to assume, as it were, an intermediate standpoint, and to select those substances of importance which are employed within the circle of observation available to us. That which has already been consigned to oblivion, or is only very rarely employed, and especially that which is no longer used by scientific medicine, deserves less attention than new drugs which may be presumed to have a valuable future. From a pharmaceutical standpoint, however, considerable interest may still be attached to many substances even though they find but little medicinal application at the present time. With regard to *Nux vomica, Santonica, Radix Belladonnæ, Gallæ,* and *Podophyllum,* by way of example, a scientifically educated pharmacist will desire to be satisfactorily informed, even when these crude substances shall have become banished from medicinal use to a much greater extent than is at present the case. In proportion as *strychnine, santonin, atropine,* etc., become of greater importance in their medicinal, or forensic and technical relations, a corresponding attention should be paid to their derivation. Neither *pepper,* nor piperine

---

[1] It occurs also as the principal title: "Pharmakognostiche Tabellen" in the fifth edition of Joh. Christ. Ebermaier's "Tabellarische Uebersichten der Kennzeichen der Aechtheit und Güte, etc., der Arzneimittel," Leipzig, 1827. The first edition of this work appeared in 1804, but whether under the above title is not known to us.

or piperidine play an important part in modern medicine, although the former constituted for centuries the most important of all spices, and still maintains, as a condiment, a prominent position in the world's commerce. Pharmacognosy would not completely fulfil its mission if such circumstances were not taken into consideration; and a similar statement might be made regarding *cacao, tea, coffee, coca, cola, guarana,* and *maté.*

Important medicinal substances become more clearly intelligible when they are compared with others which may in themselves be insignificant; and herein also lies both a demand and a justification for pharmacognosy to occasionally extend its domain in an apparently unnecessary manner. Thus the so-called *spurious Cinchona barks* for a long time presented only subordinate interest, but the consideration of their structure was admirably adapted for the demonstration of the varying peculiarities of those barks which alone furnished quinine and the allied alkaloids until the *Cinchona cuprea* appeared in the market (since 1880), and proved that these bases are by no means confined to the *true Cinchona barks.*

Hence there are various considerations which are liable to extend the compass of pharmacognosy. On the other hand, there are cases where substances which apparently belong here may be permitted to remain unconsidered. This may occur in such cases, for instance, where chemistry alone is capable of affording an exhaustive characterization. In the fats, wax, volatile oils, and the several varieties of sugar, the purely chemical properties are of such eminent importance that pharmacognosy can only in exceptional cases find motives to complete their characteristics in other directions. This task must be relegated to those departments which concern themselves with the chemico-technical or commercial knowledge of such products, for even the latter has for some years past likewise experienced the most substantial advancement.[1] The latter has been materially aided by the

[1] See Wiesner, " Die Rohstoffe des Pflanzenreiches," Leipzig, 1873, and the collective work begun in 1882 by Benedikt and eight associates, " Allgemeine Waaren- und Rohstofflehre," Cassel and Berlin, Fischer. The fifth volume represents " Die Nahrungs- und Genussmittel aus dem

earlier development of pharmacognosy, while, inversely, pharmacognosy receives at the present time many impulses from commercial relations, as an example of which may be quoted the fact that it has to derive statistical and other information relating to important drugs from consular reports. An abundant supply of such information is especially offered in the semi-annual reports of the firm of Gehe & Co., of Dresden, which may regularly be obtained from the booksellers.

Only such things fall within the sphere of pharmacognostical consideration as cannot, for our purpose, be sufficiently investigated by a single science. With regard to many leaves, flowers, seeds, and fruits, it may indeed be contended that botany is capable of affording a sufficient description ; it is, however, readily to be observed that pharmaceutical requirements demand the consideration of other than purely botanical properties. The changes which occur on drying, chemical qualities, commercial relations, and historical facts are all equally worth knowing, and call forth the discriminating and classifying ability of pharmacognosy.

It must, however, be admitted that the limitation and treatment of objects of pharmacognostical study are rather arbitrary. Pharmacognosy is by no means a branch of knowledge with sharply defined boundaries, and herein, in fact, lies the nature, and probably also a special charm of the science, that it enlists the aid of several disciplines for the single purpose of acquiring a more thorough knowledge of the crude substances employed as medicinal agents, or of parts of plants or products which are otherwise important from the standpoint of pharmacy.

Pflanzenreich," by T. F. Hanausek, 1884, pp. 485. With one hundred woodcuts.

In most cases, the following points prominently present themselves:

I. Naming the plant (or the animal) from which the substance is derived.

Here it happens not infrequently that a consideration of the synonyms is indispensable in order to avoid misunderstanding; for if we revert no farther back than to the time of Linné, we sometimes meet with plants which, since then, have been variously named by botanists. Thus, for instance, *Hagenia abyssinica* Willdenow (1790), *Bankesia abyssinica* Bruce (1799), and *Brayera anthelmintica* Kunth (1824), designate the tree which furnishes us the *koosso*. Linné, in 1737, represented the clove tree under the name of *Caryophyllus aromaticus,* but Thunberg, in 1788, named it *Eugenia caryophyllata,* properly attaching it to the genus *Eugenia,* which had existed since the year 1729. Many examples of this character are to be found in the families of the Umbelliferæ, Compositæ, and Labiatæ. Since the end of the preceding century, the mother-plant of the *calumba root* has received half a dozen, and the *sabadilla plant* four names. On the other hand, the same name has occasionally been bestowed upon different plants. Thus Linné's *Croton Cascarilla* is a different shrub from that so named by Bennett, and the *Croton Eluteria* of the latter, which furnishes the *cascarilla bark,* is not the same tree as that intended by Swartz under the same name. A similar condition exists with reference to *Cassia lanceolata,* under which name Nectoux understood the present *Cassia acutifolia,* Delile, Wight, and Arnott the *C. angustifolia* of Vahl, while Forskal's *C. lanceolata* is identical with *C. Sophera* L. Pliny had already made a distinction between the *Norway Spruce Fir*,[1] *Picea*, and the *Silver Fir*,[2] *Abies*; but Linné, in 1753, transferred the name *Pinus Abies* to the first-mentioned, which, in 1771, was again reversed by Duroi. The tree which furnishes *kamala* has received in the course of time the following names: *Croton philippense* Lamarck (1786), *Echinus philippensis* Loureiro (1790), *Rottlera tinctoria* Roxburgh (1798), *Mallotus philippinensis* Müller Argov. (1862), and *Echinus philippinensis* Baillon (1865).

Of some few drugs the mother-plants have not yet been ascertained with perfect certainty. For example, we do not know precisely whether *Rheum officinale* Baillon furnishes most or all of the commercial *Radix Rhei*. Equally unsatisfactory is our knowledge of the plants which afford *asafœtida, sarsaparilla, olibanum, elemi, copaiva balsam, galbanum, ignatia seeds*, and *Levant soap-root*. The origin of *tragacanth*, of Asiatic *salep*, and of *condurango bark* has also not yet been determined with sufficient accuracy.

II. **Geographical distribution** of those plants which are of importance in pharmacy.

However little of striking significance this subject may present from a practical point of view, it highly merits consideration in a scientific treatment of the topic. The representation of a pharmaceutically important plant is only then truly satisfactory when we are also informed regarding its habitat or the extent of its cultivation. The sources from which this knowledge is to be obtained are, in the first place, the floras of individual countries, works on plant geography,[3] and narratives of travel. Since many officinal plants are among those that have been longest employed by man, or are distributed over large tracts of territory, or are at least generally known, they find incidental con-

---

[1] *Fichte* or *Rothtanne* of the Germans.

[2] *Weisstanne* or *Edeltanne* of the Germans.

[3] A. De Candolle : " Géographie botanique raisonnée," 1855, also " Origine des Plantes cultivées," 1883. Grisebach : " Die Vegetation der Erde nach ihrer klimatischen Anordnung," 1872,—second edition, two vols., 1884.

sideration in the comprehensive works just mentioned. From a narrower, pharmaceutical standpoint, an illustrative representation of the distribution of officinal plants throughout the world has been undertaken by inscribing them upon a planisphere.[1] A survey of the respective plants is thus obtained, which, however, as may readily be conceived, has nothing to do with plant geography. The distribution or, more properly speaking, the selection of these plants can, at most, only in so far be considered as having taken place in a normal manner, as it goes hand in hand with the course of civilization. India, Persia, and the Mediterranean region, the primeval seats of civilization, have, therefore, the preponderating number of officinal plants to exhibit, Australia not a single one; the entire, enormous Arctic region perhaps only *Polyporus officinalis* and *Laminaria*, to which, if desired, *Cetraria islandica* might be added. A further objection to the graphic representation of the distribution of officinal plants may also be found in this, that the districts of production are mostly much more confined than the geographic area of the plants, because so many drugs, on account of their limited use, can play no considerable part in the wholesale trade. *Cetraria islandica*, for instance, is collected for the drug trade in the mountains of Central Europe and the Alps, and not in the far North. Finally, the distribution of useful plants is also dependent upon certain influences of cultivation.

[1] Barber: "The Pharmaceutical or Medico-botanical Map of the World," London, 1868. The same map is offered also by Fristedt, "Pharmakognostik Charta öfver jorden," Upsala, 1870, as likewise in the appendix to his "Lärobok i organisk Pharmacologi," Upsala, 1873. The map projected by Schelenz, in Archiv der Pharm., Band 208 (1876), suffers from being on altogether too small a scale, although it considers only the plants of the Pharmacopœa Germanica. This fault is avoided by Oudemans in his "Handleiding tot de Pharmacognosie," Amsterdam, 1880, since he dedicates a special map to each of the five divisions of the earth. If it is desired to go still further, then the distribution of each individual plant must be brought upon a special map. This has, for instance, already been accomplished by Lloyd for *Hydrastis canadensis* in his "Drugs and Medicines of North America," Vol. I., No. 3, p. 82 ; also in the Pharm. Rundschau, New York, p. 237.

III. **Cultivation** of officinal plants for medicinal purposes, or also for industrial applications.

The *Cinchonas* of the East Indies, Jamaica, and other countries; the *poppy* in Asia Minor, Bengal, and Malwa; the *tobacco* which is cultivated in all temperate and warm countries; the *tea plant* in Assam; the *peppermint* in Michigan; the *cinnamon* in Southern China and on the island of Ceylon; the *liquorice* in Calabria, Spain, and Moravia; the various species of *Citrus* (agrumi) in the Mediterranean region, California, and the West Indies; the *coca* in Bolivia and Peru; and the *cacao*, extending from Mexico to Brazil, are examples of the transplantation and extensive cultivation of such useful plants, without mentioning at all the *cotton plant*, the *Eucalyptus*, the *sugar-cane*, and the *sugar-beet*. In a less extensive amount, *Althæa*, *Angelica*, *Levisticum*, *fennel*, and *anise* are cultivated in Thuringia, and *caraway* near Halle; the latter two to a greater extent in Central Russia. Furthermore, the *manna-ash* in Sicily; the *rose* in Kisanlik on the Balkan Mountains and in Southern England: *peppermint* in the same locality; and *Lobelia* in the State of New York. Finally the extended cultivation of odorous flowers, near Grasse in Provence,[1] which are, however, more largely employed in the art of perfumery.

The English Government, through the prudent management of the large botanical gardens at Kew, near London, has provided for the distribution of important useful plants in India and the Colonies. To these endeavors is due the establishment of the *Cinchonas* in India, while recently the same has been accomplished with the *calumba plant*, *ipecacuanha*, *jalap*, and the trees which afford *caoutchouc*, *gutta-percha*, *copaiva balsam*, and *cinnamon*.[2] The results are still to be awaited, as are likewise those of similar endeavors on the part of the State Department of the United States.

IV. **Collection and Preparation.**—The section which fol-

[1] Flückiger, Archiv der Pharm., 222 (1884), p. 468.
[2] Compare Flückiger, "Ueber den chinesischen Zimmt," in Arch. der Pharm., 220 (1882), p. 835. Further, Brockmeier, "Ueber den Einfluss der englischen Weltherrschaft auf die Verbreitung wichtiger Culturgewächse, namentlich in Indien." Dissertation, Marburg, 1884, pp. 56.

lows later on (p. 139), relating to the liquid contents of the plant cell, shows how great are the changes produced in the plant, simply through the process of drying. It is evident that these changes have hitherto not been, by far, sufficiently considered.[1] They may often be surmised through changes in odor, as, for instance, in the case of *coriander*, the underground portions of *Orchis* (*salep*), *Iris* (*orris root*), *Veratrum* and *Aconitum*. Occasionally the drugs assume, upon drying, other colors.

It is of special importance to determine accurately the proper *time of collection* of each plant, or of its officinal parts; for during the life of a plant it does not at all times contain the active principles in equal amount—indeed, in many plants, certain constituents are, at some periods, entirely wanting. The time of collection is to be so chosen that the maximum amount of the desired substances is obtained. Quite independent, however, of the impossibility of insuring this by watching the collectors, it must also be admitted that our scientific knowledge of these conditions is still altogether too fragmentary. In the case of *Folia Digitalis*, *Folia Hyoscyami*, *Fructus Conii*, *Tuber Colchici*, *Rhizoma Filicis*, and some few other crude vegetable substances, we are, however, in this respect well informed.

Digitalis leaves are weaker in active constituents before the period of flowering than afterward, consequently the leaves of the first year are to be entirely rejected. In the case of *Hyoscyamus*, the leaves of the second year's growth are likewise preferred, at least in England. Schroff, in 1870, showed that *Fructus Conii* contains the largest amount of coniine immediately before the period of ripening. To the same investigator we are indebted for the proof that *Tuber Colchici* is only active at the flowering period of the plant. *Rhizoma Filicis*, according to all experience, should only be collected in late summer. The absolute age of the respective parts also comes often into consideration. Thus *Radix Belladonnæ* of the second or third year

[1] Experiments worthy of notice relating to this subject have already been made by Schoonbrodt. See Jahresbericht der Pharm., 1869, p. 20.

is richer in atropine than that of seven years' or still older growth, which is probably chiefly caused by the fact that this alkaloid is mainly contained in the bark, and in older roots the latter is relatively less in amount than in younger; the constituents of *belladonna leaves* are not so variable in amount.[1] The fact that many fruits and seeds before ripening contain starch, and afterwards more sugar, oil, and other substances, will be explained further on when we shall consider the subject of amylum. In the juice of *Ecballium Elaterium* Richard, elaterin occurs abundantly in July, but in September this powerfully drastic, crystallizable body is wanting therein.[2] *Pepper, cubebs*, and *cloves* are richer in volatile oil before ripening; the *Cinchona barks* may occasionally be poor in quinine, or even contain none at all.

It is self-evident that the quality of the soil must also have some influence upon the chemical development of the plant. *Valerian root* grown in dry localities is richer in volatile oil than that in moist ground, and in the case of *Taraxacum*, the root, from a chemical point of view, shows great variations, according to locality and the time of year. The red *Flores Malvæ* become blue, and *gentian root*, which in the interior is purely white, becomes colored yellowish-brown, unless the water is abstracted from it with the most extreme care. *Flores Rhoeados* do not even then retain their color, while, on the other hand, the remarkable brown coloration of *Flores Verbasci* may easily be avoided. The *Cinchona barks* also assume, during the process of drying, another color.[3]

Still other changes occur when the drugs are scalded, or are

[1] Lefort, Journ. de Pharm. et de Chim., XV. (1872), pp. 268, 421; Gerrard, Pharm. Journ., XV. (1884). p. 153. Compare further, Dragendorff, "Chemische Werthbestimmung stark wirkender Droguen," St. Petersburg, 1874.

[2] Köhler in Buchner's Repertorium der Pharm., XVIII. (1869), p. 590.

[3] The changes which the green coloring-matter of plants undergo upon drying, and the ways and means to be resorted to in order to reduce these changes to the smallest compass, have been thoroughly elucidated by Tschirch: "Einige practische Ergebnisse meiner Untersuchungen über das Chlorophyll der Pflanzen," in Archiv der Pharm., 222 (1884):

-dried over an open fire, as with *salep, jalap, curcuma,* and many Indian *aconite tubers.* This treatment manifests itself particularly by the starch assuming a pasty form. The leaves of *tea* and *maté,* which are exposed to a light roasting process, suffer still more significant chemical changes.

Marshmallow root, rhubarb, calamus, ginger; and *pepper* (white) are pared, in order to give them a better appearance, which, however, in the three latter cases is partly effected at the expense of the volatile oil. As a judicious form of treatment, on the contrary, is to be designated the so-called sweating process of the *cacao;* the slight fermentation which is thereby produced destroys a bitter principle.

V. **Commercial Relations.**—A very limited number of the medicinal substances coming under consideration here claim a position in the world's commerce, and play a conspicuous part therein. In this category, *opium* should be placed in the first line, although really only about that portion of it which does not find medicinal application, viz.: the East Indian. Of the same rank are the *Cinchona barks. Pepper, tea, cacao, tobacco, sugar,* and *maté,* regarded as luxuries, and counted among the most important products, are scarcely to be considered as drugs, which is more truly the case with *ginger.* Such drugs as are also technically employed, are usually of much greater importance in the latter respect. *Catechu, colophonium, dammar, dye-woods (logwood, Brazil-wood, etc.), galls, gum arabic, gutta-percha, turpentine,* etc., are employed, for example, in various industries in infinitely larger amounts than in pharmacy. Even for culinary purposes spices are much more largely employed than in pharmacy, notwithstanding the departure of modern taste from the former custom, to strongly season articles of food.

The most valuable information in this department, viz., statistical records, are especially to be found in the voluminous publications (Blue-books) of the English Government, and, recently also, to an increasing extent, in those of the United States. Furthermore, in the German "Handels-Archiv," and in official tables relating to the commerce of France and other countries.

VI. **Description** of the drug itself according to external characteristics, the odor and taste, and, in the case of liquids, also the specific gravity. Drugs provided with organic struc ture present many more points of observation for their examination and description.

VII. The determination of the parts which here come under consideration, according to their **organological importance.** The general considerations relating to this subject are treated of in a subsequent morphological section.

VIII. **Microscopic structure** of the drugs of organic construction.

To this subject special sections of the present work are dedicated. Among those drugs in which development does not proceed from the activity of cell formation in the organism, or is not regulated in accordance with morphological laws, there are some (as for example: *Aloe, Balsamum tolutanum, Benzoinum, Catechu, Chrysarobinum, Elemi, Opium, Styrax,* and *Terebinthina communis*) which, nevertheless, with reference to their crystalline constituents, call for microscopic examination, in the course of which polarized light renders essential service, because these crystals, on account of their double refraction, appear much more distinct under the polarizing microscope than when observed in ordinary light.

IX. **Chemical constituents.**—The enumeration and brief characterization of these principles essentially belongs to the functions of pharmacognosy. Under this heading are to be considered not only the principles peculiar to certain drugs, but also the more commonly distributed constituents of plants. Though their isolation and quantitative estimation forms the object of chemical analysis,[1] yet with the aid of micro-chemical reagents (see the concluding chapter of this book) it is possible to arrive in a short time at a series of valuable conclusions regarding individual constituents. This method of procedure especially affords information in many cases regarding the location of important constituents in the tissue.

[1] G. Dragendorff : "Die qualitative und quantitative Analyse von Pflanzen und Pflanzentheilen." Göttingen, 1882 (see below, page 49).

X. **Substitutions and Adulterations.**—An acquaintance, based on the preceding considerations, with drugs possessing an organic structure at once affords the means for their examination. When the structure of the respective substances has been rendered undiscernible in consequence of a fine degree of pulverization, the task is more difficult. In such cases, the aid of chemical analysis must especially be sought. This is rendered still more necessary when liquid drugs of variable composition are under consideration, as, for instance, *Copaiva balsam*, *Peru balsam*, *Storax*, and *Turpentine*.

XI. **History.**—The knowledge of medicinal substances would remain incomplete if their history did not also receive consideration. The investigation should extend to the time and place of first acquaintance with the mother-plant, to the time of the first medicinal employment of each single drug, and to its importance in the world's commerce. Outside of the very narrow domain of pharmacy, the relations of drugs to agriculture, to domestic economy, and to various industries, should also be permitted to be indicated, in order to illustrate the part played by them among existing commodities.

A thorough historical representation of pharmacognosy in this sense is still wanting; the preparatory labors that have as yet been brought to light admit of the following preliminary survey.[1]

1. The earliest application of the products of organic nature for medicinal purposes, as also for purposes of fumigation, refers to those countries where a higher intellectual life first became developed. In Egypt, numerous monuments of the earliest times have been preserved, which prove an acquaintance with a number of drugs at a very remote period of antiquity. Illustrative representations on temple walls, which originated in the seventeenth century B. C., inform us of Egyptian sea voyages to the provinces of northeastern Africa and Arabia, which were in part undertaken in order to procure *gum arabic*, *frankincense*,

[1] For appropriate descriptions we are also indebted to Schär, " Die ältesten Heilmittel aus dem Orient." Schaffhausen, 1877, pp. 24, and " Aus der Geschichte der Gifte." Basel, 1883, pp. 48.

2

and *myrrh*. It is, indeed, possible that through these primeval trade relations spices and medicinal substances found their way to Egypt[1] even from southern and eastern Asia. In the inscriptions of temples,[2] and in the rolls of papyrus, frequent reference is made to such things, the correct interpretation of which is still in part uncertain; only a few of them have actually been brought to light from burial-vaults.[3] Especially in the oft-recurring, but not corresponding recipes for *Kyphi*, a medicinal compound which served for manifold purposes, there occur a large number of drugs, as, for example, *mastic, cardamom, curcuma, ladanum* (resin of *Cistus ladaniferus*) and *fenugreek*.[4] It will only be after a still more extended investigation of Egyptian antiquities that the extent of the knowledge in question among the ancient Egyptians can be established. Through their highly developed knowledge of agriculture they also engaged in the cultivation of several of the plants which here come under consideration, as, for example, *coriander, fenugreek, flax, poppy, ricinus,* and *sesame*.[5]

[1] Dümichen, " Die Flotte einer ägyptischen Königin," 1868, and "Historische Inschriften," 1869. Mariette-Bey, " Deir-el-Bahari," 1877. Flückiger, " Pharmakognosie," second edition, pp. 6, 35, 41, 560, 935. Schumann, "Zimmtländer," Gotha, 1884 (Supplement No. 73 to Petermann's Mittheilungen).

[2] Dümichen, " Das Salbölrecept aus dem Laboratorium des Edfu-Tempels." Zeitschr. für ägypt. Sprache und Alterthumskunde, December, 1879.

[3] Braun (Ascherson and Magnus), " Ueber die im Museum zu Berlin aufbewahrten Pflanzenreste aus altägyptischen Gräbern." Zeitschrift für Ethnologie, IX., Berlin, 1877, pp. 289 to 310, and Schweinfurth, " Ueber Pflanzenreste aus altägyptischen Gräbern." Ber. d. deutsch. botan. Ges., III. (1884), p. 351.

[4] Lepsius, in the Zeitschr. für ägypt. Sprache und Alterthumskunde, October, 1874, p. 106: " Kyphirecept aus dem Papyrus Ebers."

[5] Compare further Thaer, " Die altägyptische Landwirthschaft," Berlin, Parey, 1881, 4to, pp. 36, with six plates. Unger, " Botanische Streifzüge auf dem Gebiete der Culturgeschichte." Sitzungsberichte der Wiener Akademie, 1857 to 1859, especially in volume 45, " Inhalt eines ägyptischen Ziegels an organischen Körpern," and volume 54 (1866): " Ein Ziegel der Dashurpyramide in Aegypten nach seinem Inhalt an organischen Einschlüssen." Schweinfurth, " Blumen-

2. The industrious trading nation of the Phœnicians,[1] and through them probably the Israelites, were as well acquainted as the Egyptians with the above-mentioned drugs, to which from the Old Testament Scriptures may still be added, *aloes, cinnamon, coriander, saffron, ginger, olive-oil, sugar,* and *pepper.*[2] In their religious ceremonies these people evidently employed aromatic substances in large amount. Drugs which at that time were exceedingly highly prized, but which for a long time past have become completely obsolete, were *Radix Costi*[3] and the *Aloeswood* of *Aquilaria Agallocha,* Roxburgh.[4] The high esteem with which these two aromatic substances were regarded from that time until the eighteenth century is scarcely intelligible to us: and in India and China it still continues undiminished.

3. The Chinese were also, without doubt, at a very early period, familiar with the medicinal substances which were native to that country, as, for instance, with *camphor*, with remedies from their animal world, and with such from the mineral kingdom. Since cinnamon was certainly exported in the earliest times, it may be presumed that foreign drugs were probably at that time also imported into China. The respective ancient literature of this country, however, has still been too little scrutinized to afford reliable information regarding these conditions.

schmuck ägyptischer Mumien," Gartenlaube, Leipzig, 1884, 628 and *loc. cit.* (note 3). [Also, particularly, Woenig, "Die Pflanzen im alten Aegypten," Leipzig, 1886. F. B. P.]

[1] Flückiger, "Pharmakognosie," pp. 119, 120, 560.

[2] The plants mentioned in the Bible have led to widely-extended discussions. It is sufficient here to name the following modern writings relating to the subject: Cultrera, "Flora biblica ovvero spiegazione delle piante menzionate nella S. Scrittura," Palermo, 1861; Ursinus, "Arboretum bibl. c. contin. hist. plant. bibl.," November, 1865; Duschak, "Zur Botanik des Talmud," Pest, 1871; Hamilton, "Botanique de la Bible," Nice, 1872; Smith, "Bible Plants, their History, a Review of the Opinions of Various Writers regarding their Identifications," London, 1878; Wilson, "The Botany of Three Historical Records (Genesis, New Testament, Assise of Weight and Measure)," Edinburgh, 1879; Fillion,"Atlas d'Histoire naturelle de la Bible," Paris, 1885, 4to, pp. 116, 112 plates.

[3] Flückiger, "Pharmakognosie," pp. 444, 906. Dymock, "Materia Medica of Western India," 1884, p. 372.

[4] Flückiger, *loc. cit.*, 195.

So much is certain, however, that the principal work of the
Chinese, the herbal *Pen t'sao kang mu,* which is, indeed of a
much later date, is based in part upon very much older sources.[1]
The highly developed condition of popular medicine of this
people,[2] which adheres so tenaciously to primeval customs, refers
to a period of very remote antiquity. For information relating
to many of the pharmaceutically important plants of China,
pharmacognosy is indebted to the great Venetian traveller,
Marco Polo (toward the close of the thirteenth century), as
also in the eighteenth century to the missions of the Jesuits in
China.[3]

That an early acquaintance with medicinal plants and
drugs existed in Japan has not yet been proved, but it may certainly be assumed, for instance, as regards *menthol,* " Hakka."

4. As to how far this is the case with regard to India has likewise not been established. Sanskrit literature possesses in
" Susruta " and " Charaka " information relating to medicinal
substances, which are probably in part of more ancient origin;
but recent investigation assigns to these writings a much less
ancient date.[4] The extent of the pharmacognostical knowledge
of Indian antiquity therefore requires more exact investigation;[5]
the application in India of many medicinal substances and products of the vegetable kingdom adapted to fumigation, such as
white sandal-wood, camphor, cinnamon, and *cardamon* may cer-

[1] Flückiger, *loc. cit.,* p. 1012.

[2] Hanbury, "Science Papers," 1876, 211 to 272. Flückiger, Archiv der
Pharmacie, 214 (1879), p. 9.

[3] The activity of this order deserves further mention in connection with the history of some other drugs, as, for instance, the *cinchona
barks, ginseng root, maté,* and *ignatia seeds.* The *sassafras tree* appears
to have been brought by Jesuits from Canada to France; the earliest information regarding the tape-worm remedy *koosso* probably likewise
originated from a member of this order. In Rome, Manila, Paris, and
South America they maintained pharmacies, which were probably always conducted by Jesuit friars.

[4] Compare Flückiger, "Pharmakognosie," p. 1020.

[5] Only a few points of guidance in this direction have as yet been
offered by Lassen's "Indische Alterthumskunde," Bonn, 1847-1852.

tainly be traced to a very remote period. It may be presumed that *musk* has also been in use there for a very long time.

5. The centuries representing the flowering period of Greek and Roman civilization considerably increased the number of medicinal substances, not only through such as were obtained from the region of the Mediterranean, but also through others of Oriental origin. Among these are especially: *Amygdalæ dulces, Bulbus Scillæ, Cantharides, Caricæ, Castoreum, Cortex Granati,*[1] *Euphorbium, Fructus Anisi, Fructus Cardamomi,*[1] *Fructus Fœniculi, Fungus Laricis, Gallæ, Herba Sabinæ, Indigo, Mastiche, Opium,*[1] *Piper longum, Radix Glycyr-rhizæ, Radix Rhei (?), Rhizoma Filicis, Rhizoma Iridis, Sandaraca, Scammonium, Semen Fœni grœci,*[1] *Semen Lini, Semen Sinapis, Succinum, Siliqua dulcis, Succus Glycyrrhizæ, Terebinthina* and *Tragacantha.*

That the number of plants employed in classical antiquity was very considerable is shown particularly in the writings of Dioscorides and Pliny,[2] to which the following centuries until the close of the European middle ages continually refer, and indeed, almost without any advancement, on their own part, of the existing knowledge. Many plants of the Italian flora which are employed medicinally have often been thoroughly considered by Roman writers on agriculture, namely, by Cato, Columella, Palladius and Varro. Columella[3] (in the years 35 to 65 A. D.) had also made observations in Spain and Syria. The principal contents of their writings, so far as they come into consideration here, are to be found compiled in a very complete and systematically arranged form in Magerstedt's " Bilder aus der römischen Landwirthschaft."[4] Hehn also, though with much more spirit

---

[1] These might have perhaps also been quoted under division 2, page 19.

[2] To the editions of Pliny mentioned in Flückiger's " Pharmakognosie,'' pp. 997 and 1,014, the new translation by Wittstein may be added.

[3] Flückiger, *loc. cit.*, p. 991, etc. Compare also, Meyer, "Geschichte der Botanik," Bd. I. and II., Königsberg, 1854, 1855.

[4] Heft IV., Sondershausen, 1861 : " Die Obstbaumzucht der Römer;" Heft V. (1862): " Der Feld-, Garten- und Gemüsebau der Römer:" VI., (1863): " Die Bienenzucht und die Bienenpflanzen der Römer." The author does not enter upon the consideration of the significance of the plant names.

and taste, incidentally touches upon this subject in his publication: "Kulturpflanzen und Hausthiere in ihrem Uebergang aus Asien nach Europa." Berlin, 4th edition, 1882. A very remarkable light is thrown upon the intercourse of the Occident with the Orient through the *Periplus of the Erythræan (Red) Sea.* This survey of the coast of the East Asiatic-Indian Ocean, which was accomplished in the first century after Christ, enumerates a number of the commodities [1] which were to be met with in those ports, and among these many of pharmaceutical interest, such as *myrrh, Sanguis Draconis, Styrax liquidus, sandal-wood, pepper, frankincense,* and *saffron.*

The continued importation of the Indian spices mentioned under division 2, page 19, is proved by a list dating from the years 176 to 180 A.D. In this the commodities are enumerated which arrived from the Red Sea, [2] and which were subject to the Roman duty at Alexandria.

It is quite natural that even at that time the art of adulteration was also applied to drugs. A single sentence from Pliny is amply sufficient for the confirmation of this fact; for, when treating of *saffron,* the well-versed Roman encyclopædist employs the expression: "adulteratur nihil æque." [3]

Even with regard to the relation of prices of some few drugs in those early times, Pliny has furnished us some information. Thus, for example, a pound of *black pepper* was estimated at 4, *white pepper* at 7, *long pepper* at 15, and *indigo* [4] (also named as

---

[1] Enumerated in Meyer's "Geschichte der Botanik," II. (1855), p. 85. Compare also, Flückiger, *loc. cit.,* p. 1,013.

[2] *Ibid.,* p. 985.

[3] Flückiger, *loc. cit.,* p. 740.

[4] Pliny, who in such matters was certainly quite inexperienced, must have had an attentive observer as a source for his statements here ("Nat. Hist.," XXXV., 27; page 470 of the edition by Littré). After a good description of *indigo (indicum),* heating of the same is recommended, in order to test the vapor and the peculiar odor: "probatur carbone; reddit enim, quod sincer est, flammam excellentis purpuræ et, dum fumat, odorem maris."

When treating of verdigris, Pliny (XXXIV., 26) also resorted to a chemical reaction, in order to recognize in it an admixture with iron;

a medicament) at 20 *denarii*. A larger number of prices is given in Diocletian's "Edictum de pretiis rerum venalium," of the year 301 A.D., which, indeed, applied more to articles of food and other indispensable necessaries of life than to spices and medicinal substances, and was also issued only for the eastern part of the Roman empire, not for Europe.[1] The edict mentions, among other things, *almonds, hemp-seed, figs, fenugreek, linseed, olives, mustard, sesame seed,* and *grapes*.

6. During the decline of antique civilization the fostering of the sciences passed into the care of the Arabs, who also seized upon the medical traditions of antiquity, and were enabled to widely extend them, through their geographical position, from the highlands of Asia and India to Spain and northern Africa, as also, through their frequent connections with India, occasionally to enlarge and expand the same. The most eminent representatives of Arabic medicine in the tenth and eleventh centuries, Alhervi, Avicenna, Mesue, Serapion, and others, enriched materia medica with some Asiatic drugs, such as *tamarinds, nux vomica, cubebs, senna leaves, rhubarb, camphor,* and *worm-seed* (?), and, through the introduction of formulas for medicinal preparations, exercised an exceedingly lasting influence upon pharmacy, even in the Occident. With regard to many drugs from distant lands, the Arabians at an early period received information through travellers, or from historical and geographical authors,[2] such as Khurdadbah, Istachri, Masudi, Idrisi, and Ibn Batuta. Ibn Alawan reported in the twelfth century upon the flourishing condition of agriculture among his people in Spain, from which also dates the cultivation of *saffron*, still continued there at the present day. The most prolific information, however, chiefly from earlier and often much older sources of Arabic literature, has been brought together by Ibn Baitar in his large encyclopædia of simple medicinal remedies and

one should employ for this purpose paper soaked in nutgall—thus probably the earliest application of test-paper !

[1] Flückiger, *loc. cit.*, p. 997. Compare further, Burckhardt, "Die Zeit Konstantins des Grossen." Leipzig, 1880, p. 62.

[2] Flückiger, *loc. cit.*, p. 107, Note 10.

foods.[1] The critical examination and often very difficult interpretation of many statements relating to this subject contained
in Arabic literature, which is constantly progressing, admits of
the hope of some very remarkable revelations in this respect.

A pharmaceutical manual of Aboul Mena, called Cohen el At-
thar (Priest and Apothecary), who lived in the thirteenth century
at Cairo, has not yet been printed.[2]

7. In the far West, these antique recollections were likewise
partly revived at the same time in civil and ecclesiastical
quarters. Thus the Emperor Charlemagne, through special
edicts, issued in the year 812, ordered the cultivation, north of
the Alps, of a series of long-known useful and medicinal plants,
among which may be mentioned :[3] *Althæa, Amygdalus, Anisum,
Coriandrum, Cydonia, Fœniculum, Iris (Gladiolus), Levisticum, Mentha, Petroselinum, Rosmarinus, Ruta, Sabina, Salvia,*
and *Sinapis.* It is worthy of note that the following useful
plants, which are indigenous to Italy or generally cultivated
there, are wanting in the " Capitulare," or the principal one of
those Imperial edicts, viz.: *Glycyrrhiza, Inula Helenium,
Lavandula, Punica Granatum,* and *Thymus vulgaris.* In the
architectural plan of the convent of St. Gall, which was drawn
in the year 820, but was not executed, the place which the
medicinal plants were to occupy in the garden was designated,
perhaps in compliance with the " Capitulare."[4]

In the library of the University of Würzburg, Germany, a
recipe for a mixture of powders, " contra omnes Febres et contra
omnia venena et omnium Serpentium morsus et contra omnes
angustias cordis et corporis," has been preserved which is perhaps
a century older than the preceding. This curious manuscript

[1]Flückiger, *loc. cit.*, p. 1,003.

[2] According to Leclerc, " Histoire de la médecine arabe," II. (Paris,
1876), p. 215, it appears very worthy of notice.

[3] The complete list is given by Pertz, " Monumenta Germaniæ his-
torica," legum tom. 1. (1835), p. 186, and also in Meyer's "Geschichte
der Botanik," III., p. 401. Compare further, Flückiger, *loc. cit.*, p.
1,005.

[4] *Ibid.*, p. 688.

is probably one of the oldest monuments of popular medicine in Germany; in it are mentioned *Pimpinella* and *Galanga*, both, to the best of our knowledge, for the first time.

8. The determining influence upon the medicine and pharmacy of the middle ages proceeded from the medical school of Salerno. Its activity extended from the ninth century until the close of the middle ages, although the school itself was continued, at least in name, until November 29th, 1811, when it was suppressed by an order of Napoleon. To the medical practitioners who taught there, and in the not very remotely located Benedictine monastery of Monte Cassino, is due the credit of having transmitted the medical art of the Arabs. Thus the stock of esteemed remedies of the Occident was increased by a number of new medicinal substances, or at least such as were previously but slightly accessible. To these belong *Ammoniacum,*[1] *Asafœtida,*[1] *Benzoinum, Camphora, Caryophylli, Cinnamomum zeylanicum, Cortex Aurantiorum, Cortex Limonum, Cubebæ* (as a spice; first used medicinally since 1813), *Folia* (or probably at first only *Siliquæ*) *Sennæ, Fructus Cocculi, Fructus Colocynthidis, Galbanum,*[1] *Herba Cannabis, Lignum Sandali, Macis, Moschus, Radix Rhei, Resina Draconis, Rhizoma Curcumæ, Rhizoma Galangæ, Rhizoma Zedoariæ, Semen Myristicæ, Semen Strychni, Styrax liquidus, Tamarindi,* and *Salep.*

The writings of the most prominent Salernitans, namely those of Constantinus Africanus, Macer Floridus, Nicolaus Præpositus, Platearius, Arnaldus de Villanova, and especially the " Regimen sanitatis Salernitanum " must be studied in order to obtain an insight into the science of medicine of the middle ages. A very complete enumeration of the *Simplicia* and *Composita* of that time is given by Renzi in a list published under the name of *Alphita,* which is presumed to belong to the thirteenth century.[2]

[1] Presumably known at an earlier period.

[2] The history of the School of Salerno, which is not yet sufficiently elucidated, has been briefly but very appropriately described by Handerson, in the publication mentioned in Flückiger's " Pharmakognosie," p. 1,018.

9. While the erudition of southern Italy interested itself in the maintenance and extension of the sciences of the Orient, the commercial republics of Italy—Venice, Amalfi, Pisa, Florence, and Genoa—were enabled through their fleets to obtain from the far East and South such commodities as were required in medicine as well as for the more refined enjoyment of life and the improvements in handicraft. Venice, which was by far the strongest of these commercial states, began already in the eleventh century to develop the elements of its incomparable splendor, and to become the central point for the traffic in drugs. Up to the end of the sixteenth century there flowed into this city, in the greatest abundance, those spices which were sought with such extreme eagerness, but the value of which, in our time, has become very greatly reduced. The greatest importance was attached to *pepper*, the commercial history of which vividly reflects this entire and exceedingly remarkable intercourse of nations, as, indeed, in the middle ages, *pepper* represented the symbol of the entire spice trade.[1]

The Levant trade of Venice, which for those times was incomparably large, also gave an impulse for the establishment, in that city, of chemical industries, such as the manufacture of sal-ammoniac, corrosive sublimate, cinnabar, soap, and glass, together with the bleaching of wax and the refining of borax (*Tinkal*, from Thibet) and of camphor.

Through the shrewd participation of the Venetians in the crusades of the twelfth and thirteenth centuries, their commerce and influence attained the highest degree of prosperity. Even the continued warlike conflicts of the Occident with the Orient must have necessarily also contributed to the knowledge and distribution of special medicinal substances in other parts of Europe. Many historians of that time, who visited Palestine,

[1] The spice dealers in various countries were known as *Piperarii*. Such a " Gild of Pepperers " existed already in London in the year 1345, and the " Society of Apothecaries " still existing there, which received its charter from James I., in the year 1617, properly dates back to those pepper dealers. Compare Pharm. Journ., XV. (1884), p. 367; also Flückiger's " Pharmakognosie," p. 867.

described, with great clearness, for instance, the *sugar-cane* and *sugar*, while others made themselves acquainted with the *agrumi* (the fruits of species of *Citrus*), with *licorice*, *dates*, *cotton*, and *cumin* (*Cuminum Cyminum*).[1]

It was only subsequent to these voyages that *sugar* became a regular article of commerce, chiefly in the hands of the Venetians. The successful cultivation of saffron in England and France during the middle ages, and the culture of *roses* in the province of Champagne was started, or at least probably received a new impulse, through the crusaders.

In Germany the great Benedictine monasteries, for instance, those of St. Gall and Fulda, formed the central points of intellectual culture, and likewise supported and distributed botanico-medical knowledge, even though they did not actually augment it. The first monastery of this in many respects highly meritorious order, which was founded by St. Benedict himself, in the year 528, at Monte Cassino, northwest of Naples, and afterward became so greatly celebrated, stood in the eleventh century in close relation to the medical school of Salerno. This school was also of influence in determining the intellectual course of St. Hildegard, who, in the year 1148, became abbess of a convent of the Benedictine order, founded at her own instigation near Bingen on the Rhine. To this personage is attributed, although not with absolute certainty, a work which is very remarkable from a pharmacognostical point of view: "Subtilitatum diversarum naturarum creaturarum libri novem," and is presumed to have been written about the year 1178. By the enumeration of a number of indigenous plants, to which some prominent characteristic, or often also the German name,[2] was occasionally added, the book of Hildegard, frequently designated as "Physica," reveals itself as a truly German production. Of incom-

---

[1] Compare the respective articles in the "Pharmacographia" and in Flückiger's "Pharmakognosie;" also Meyer, "Geschichte der Botanik," IV., p. 110; Heyd, "Levantehandel," II., p. 670.

[2] Compare also Pritzel and Jessen, "Die deutschen Volksnamen der Pflanzen. Aus allen Mundarten and Zeiten zusammengestellt." Hannover, 1882.

parably greater importance, however, is the Dominican monk, known as Albertus Magnus, who from the year 1260 to 1280 was Bishop of Regensburg (Ratisbon). In his books "De Vegetabilibus," numerous medicinal plants and drugs have received for the most part very intelligent treatment, though this is largely based upon other sources of information, including Arabian.

The history of indigenous, and probably also foreign medicinal plants may be further pursued with the aid of the dictionaries and glossaries of the middle ages. Compilations of this kind[1] were also specially made use of in attempting, quite in accordance with mediæval custom, to identify in the indigenous flora plants of classical antiquity. From a mercantile point of view, very valuable information is afforded by the ordinances and lists of the departments of customs[2] of that time, as also especially by the trade-books of the Venetians and Florentines.[3] The extraordinary importance of the commercial intercourse with the East during the middle ages has finally been described in an exhaustive and captivating manner by Heyd,[4] who not only represents the political and economical sides of the subject, but also gives detailed accounts of the most important objects of that remarkable commercial era, namely: *aloes, aloes-wood, ambergris, balm of Gilead, benzoin, Brazilwood, camphor, cloves, frankincense, galanga, galls, ginger, indigo, kermes* (a coloring insect afterwards supplanted by *cochineal*), *lac, mace, musk, nutmegs, pearls* (both as a medicament and ornament), *pepper, precious stones* (quite a series of these were employed as medicinal agents), *rhubarb, saffron, sandalwood,* and *tragacanth.* All that the civil and ecclesiastical travellers and historians of those centuries were able to report relating to these articles has been made use of in the proper place by Heyd, in his careful examinations.

[1] Flückiger, "Pharmakognosie," pp. 107, 330, 642, 688, 713, 892, 898.
[2] *Ibid.,* pp. 781, 983.
[3] *Ibid.,* pp. 1,011, 1,012.
[4] "Geschichte des Levantehandels im Mittelalter," 2 vols., Stuttgart, 1879.

The precious goods from India were obliged, as a rule, to take their way through the Red Sea and through the domain of the Egyptian sultans, for which reason the Italian merchants of the middle ages were compelled to bestow the greatest attention upon their relations to those rulers. The ambassadors of the latter to the doges of Venice, to a Venetian queen of Cyprus, and to Lorenzo de' Medici in Florence, brought rare drugs of the Orient to Europe, as, for instance., *aloes-wood*,[1] *civet, Mecca balsam*,[2] *myrobalans*,[3] *opium*, and *sugar*, the latter of which was at that time, or in the second half of the fifteenth century, still a rare article. Thus in the year 1461 *benzoin* was brought for the first time to Venice.[4]

It was only in case of necessity that the Italians permitted their goods from the Orient to take the much longer route through the Persian Gulf, or even through Central Asia to the Black Sea, instead of through Egypt.

The intercourse with the Levant consisted quite especially in the importation of numerous Asiatic products; there were but few articles which the Italians, inhabitants of Southern France, and Catalonians, for example, had to ship to Alexandria. The Venetian statesman, Marino Sanudo,[5] in the year 1307, mentioned as such articles *almonds, honey, hazelnuts, mastic*, and *saffron*.

The most prominent Florentine houses also conducted the Levant trade with consummate mastery. Thus, for example, in the first half of the fourteenth century, the large trade association which bore the name of the leading house, Bardi in Florence. It was in their service that Pegolotti, about the year 1340, wrote that very remarkable trade-book, "Pratica della mercatura,"[6] which gives the most instructive informatio nregarding the commercial relations, coins, weights, measures, and

---

[1] Flückiger, "Pharmakognosie," p. 195, and this work, p. 19.
[2] *Ibid.*, p. 130.
[3] *Ibid.*, p. 244.
[4] *Ibid.*, p. 113.
[5] Flückiger, "Pharmakognosie," p. 1,018.
[6] Compare further Heyd, loc. cit., i., p. xiii.

trade products of that time, and is far superior to similar publications of a later date.

A less comprehensive, but nevertheless remarkable book, is that of the Venetian Pasi or Paxi, " Taripha," [1] of the year 1503.

10. Sources of information are not wanting for judging of the supply of commodities, and also of preparations which were to be met with in the fifteenth century in German pharmacies. The " Frankfurter Liste " and the " Nördlinger Register " exhibit quite a series of drugs [2] which, at that time, were actually kept in stock; and the inventory of a pharmacy in Dijon of the year 1439 is also known.[3]

The conquest of Constantinople by the Turks on the 29th of May, 1453, as also the discovery of a sea passage to India in May, 1498, are events which, by its causes and results, led to the gradual but almost complete downfall of the Levant trade, and especially to the exhaustion of the resources of Venice. The discovery of America completed the enormous revolution. The age of great discoveries could at first gain no greater number of new, important drugs from the vegetable kingdom of Asia, which had already been the source of supply for so long a period. The discovery of a sea passage around the Cape of Good Hope, however, resulted in a much more abundant importation of familiar commodities of that character. Furthermore, the scientific world now finally received more exact intelligence relating to the celebrated products of India. For this information, thanks are primarily due to a Portuguese physician, Garcia de Orta (Garcias ab Horto), who resided for thirty years in Goa, India. His discourses upon Indian drugs, which appeared at Goa in the year 1563, form a highly agreeable contrast to the mostly confused and altogether too meagre notices of the Arabs and of Marco Polo, the great traveller, whose reports are otherwise so valuable.

The most substantial advancement of pharmacognostical

[1] Flückiger, " Pharmakognosie," p. 1,011.
[2] See page 3, note 1.
[3] Schweizerische Wochenschrift für Pharmacie, 1873, Nos. 6, 7, 8.

knowledge at that time in Europe proceeded from Clusius, who, after obtaining a broad preliminary education, was engaged in botanical pursuits, principally at Vienna, and finally, until the year 1609, at Leyden. As early as the year 1567, this distinguished man published a Latin translation of Garcia's "Coloquios" under the title "Aromatum et simplicium aliquot medicamentorum apud Indos nascentium historia." Clusius freed the work of the Portuguese of its labored diction, omitted many useless and speculative additions of the author, and, on the other hand, added valuable notes comprising personal experiences and observations. Clusius was already acquainted with the *cola nut, star-anise, gamboge, winter's bark, sabadilla,* and *vanilla.*

The most instructive and abundant information relating to the Indian flora, combined with illustrations of long-known, celebrated medicinal plants, is presented (though, indeed, at a much later period) in the twelve folio volumes of Rheede's "Hortus indicus malabaricus," which appeared at Amsterdam between the years 1678 and 1703, as also in the "Herbarium amboinense," by Rumphius (Rumpf), in six volumes, Amsterdam, 1741–1755. These grand achievements of the Dutch have been followed in modern times by the magnificent works of the English botanists.[1]

Less laudable, even though it is intelligible when considered in the light of that period, appears the commercial policy of the Dutch in the seventeenth and eighteenth centuries, through which they were able to retain for themselves the exclusive possession of Indian spices, particularly of *nutmegs, cloves, pepper,* and *cinnamon,* occasionally even advancing the prices by the destruction of these articles when the stock had accumulated to too great an extent.

In the first half of the sixteenth century the knowledge of crude medicinal substances had meanwhile been much promoted in Germany by the talented teacher Valerius Cordus (1515–1544), after whose premature death his most important writings were first published, in the year 1561, by his friend and associate

---

[1] See Archiv der Pharm., 222 (1884), p. 249.

Conrad Gesner.   Cordus, in his annotations to Dioscorides and in the "Historiæ Stirpium (s. Plantarum)," in opposition to those of his predecessors who were involved in the study of antiquity, recorded good individual observations, and described the drugs much more carefully, particularly according to his personal inspection; thus, for example, *Nux vomica, Cocculus Indicus*, and *Lignum Guaiaci.*[1]   Cordus has rendered valuable service to practical pharmacy by the compilation of a dispensatory[2] in the years 1542 and 1543, by request of the authorities of the city of Nuremberg.

Under the title of "Horti Germaniæ," Gesner, in the year 1560, gave interesting accounts of medicinal and useful plants, or otherwise noteworthy species, which he or his friends, mostly apothecaries,[3] cultivated in Germany.   Gesner permitted this publication to appear as an appendix to that of his friend Cordus.

At a still earlier period, the flora of the surrounding country had been brought into requisition, but without much critical discrimination, as was the case with the widely diffused, popular book, "Hortus Sanitatis."   The same degree of consideration was accorded to the curious work on distillation by the surgeon Hieronymus Brunschwig, of Strassburg, first published there in the year 1500, and which gave the crudest instruction relating

[1] A still earlier monographic publication relating to *Lignum Guaiaci*, which, for those times, is very worthy of notice, and written in excellent Latin, likewise emanates from the pen of a prominent German, the Knight Ulrich von Hutten: "Vlrichi de Hutten Eq. de Guaiaci medicina et morbo gallico liber vnus." 4to, 26 chapters, unpaged. The first of the numerous editions, printed in the year 1519 in Schaffer's house at Mayence, bears at the end of the work a woodcut likeness of the author. This publication, by its accuracy and instructiveness, excels most of its contemporaries in the description of new medicinal substances. Hutten describes the habitus of the tree, the wood, bark, and resin, and lastly the applications.

[2] Flückiger, "Pharmakognosie," p. 994.

[3] Among foreign apothecaries Gesner mentions also the meritorious Peter Coudenberg, of Antwerp (Flückiger, loc. cit., p. 458); also Follietus, of Vevey, on Lake Geneva, Switzerland.

to the preparation of distilled waters, whereby all conceivable herbs, and things of an entirely different nature, were brought together in a senseless manner. Here and there Brunschwig inserted a notice worthy of consideration, relating to a plant, for instance, one from which it may be concluded that at the end of the fifteenth century *lavender* was cultivated in Germany, and *anise* near Strassburg. Brunschwig's book corresponded, moreover, so much to the views of that time, that in the year 1597, an Augsburg physician, Adolf Occo, introduced one hundred and forty different *Aquæ destillatæ* into the " Pharmacopœa Augustana."

Subsequent to the year 1531, Otto Brunfels had numerous figures of plants prepared at Strassburg, some of which are very good, and are still worthy of notice as the earliest good examples of the application of woodcuts to botanical purposes. His descriptions are, indeed, of an incomparably lower order than those contained in the herbals of his direct successors, Cordus (see p. 32), Bock (Tragus), Fuchs and Gesner. All these German " fathers of botany" of the sixteenth century, though not exclusively physicians, have also, through their writings, furnished pharmacy with a more exact knowledge respecting the indigenous medicinal plants, a large number of which were, at that time, kept at the pharmacies. It was also at this period of general reformation that Brunfels (in his " Reformation der Apotheken ") brought forward a list of drugs, prepared without much judgment, but containing those which appeared to him worthy of commendation, under the title " Ein gemeyne besetzung einer Apotheken, von Simplicibus."

11. To the discovery and settlement of America we are indebted for a number of substances already in use by the people of that country, which soon found their way to Spain, Portugal, and other parts of Europe. The writers of those two countries, Gonzalo Fernandez (Oviedo), Monardes, and Hernandez, who were the first to devote attention to the natural products of the new world, also began to describe medicinal plants more accurately than had been done before.

Thus America gradually furnished: *Balsamum Copaivæ,*

3

Bals. peruvianum, Bals. tolutanum, Cascarilla, Cinchona, Elemi, Folia Nicotianæ, Fructus Capsici, Fr. Pimentæ (Amomi), Fr. Sabadillæ, Lignum Hæmatoxyli, Lig. Fernambuci (Brazil wood), *Lig. Guaiaci, Lig. Quassiæ, Lig. Sassafras, Radix Ipecacuanhæ, Rad. Sarsaparillæ, Rad. Senegæ, Rad. Serpentariæ, Resina Guaiaci, Semen Cacao, Sem. Sabadillæ, Tuber Jalapæ,* and *Vanilla.*

At a later period (1636–1641), the German geographer and astronomer Marcgraf, and the Dutch physician Piso, were jointly commissioned by the government of the Netherlands, to promote the knowledge of Brazilian medicinal plants. Previous to this they had already furnished descriptions of Copaiva, Elemi, Jaborandi, Ipecacuanha, Matico and Tapioca.

12. The re-awakening of the sciences at the beginning of the new era also led to the gradual foundation of scientific botany, and in the seventeenth and eighteenth centuries to the medicinal employment of a large number of plants of Central Europe, as well as of some few drugs from other parts of the world, which then arrived in Europe for the first time. Of the long series of plants and substances which are properly to be enumerated in this connection, there may be mentioned: *Catechu, Cortex Frangulæ, Flores Arnicæ, Flores Chamomillæ, Folia Aconiti, Fol. Digitalis, Fol. Laurocerasi, Fol. Menthæ piperitæ, Fol. Toxicodendri, Fol. Uvæ ursi, Herba Chenopodii ambrosioidis, Herba Cochleariæ, Herba Conii, Herba Hyoscyami, Herba Lobeliæ, Fructus Anisi stellati, Gummi-resina Gambogia, Kino, Lactucarium, Lichen islandicus, Lycopodium, Oleum Cajuputi, Ol. Rosæ,* and other distilled oils, *Radix Calumbæ, Rhizoma Caricis, Rhiz. Filicis, Saccharum lactis, Secale cornutum, Tuber Chinæ* and *Tuber Colchici.*

13. In the cities of Germany, the stock, and especially the prices of drugs in the pharmacies, were regulated by official ordinances, while pharmacies were also openly conducted on the part of the magistrates, or leased by them. In no other country at that time were the affairs of pharmacy the subject of such thorough official attention. The manuscripts relating hereto admit of a complete insight into the store-rooms and laboratories

of the pharmacies from the beginning of the sixteenth century. Such, for example, are the inventories of the senatorial pharmacy at Brunswick (Braunschweig) of the years 1518 to 1658,[1] the catalogues of several other pharmacies, and a large number of price-lists which were issued in the course of the sixteenth, seventeenth, and eighteenth centuries in all German countries for the use of the apothecaries. It was rarely, however, that the latter were mentioned as participants in the official control, reference being usually made only to physicians and officials.

By the aid of these manuscripts, many of which still remain buried in archives, the introduction and gradual distribution of many drugs may be traced, as, for example, some of those derived from America. These documents also afford positive information concerning the extensive cultivation of certain medicinal plants, which has long since been abandoned, as, *e. g.*, the *Angelica* near Freiburg, in the old district of Breisgau, Baden; *Licorice,* near Bamberg; *Saffron,* in England, Germany, and Austria, and *Cassia obovata* in Tuscany.

An insight is also afforded into the obscure department of adulterations, to which drugs, as well as articles of food and luxuries, were subjected in the middle ages, no less than at any other time (see p. 22). The magistrates of German and Italian cities resorted to the most extreme police measures, even punishment by death,[2] for such offences, and authorized physicians to watch the apothecaries. In the ordinances accompanying the above-mentioned price-lists, and in special manuscripts, these regulations are quoted at great length.

The unfortunate apothecary Zanoni de' Rossi, of Venice, in the year 1402, when preparing the highly celebrated "Theriac," was detected in the omission of *Rhubarb, Amomum, Opoponax* ("Pharmacographia," p. 327) and *Saffron,* even substituting

---

[1] For a copy of this, one of us (F.) is indebted to the apothecary Dr. Grote, in Brunswick. Compare also the notice of the latter in the Archiv der Pharm., 221 (1883), p. 417.

[2] Flückiger, "Pharmakognosie," p. 740. Compare further the thorough work by Elben, "Zur Lehre von der Waarenfälschung." A Tübingen dissertation, 1881.

for the latter *Safflower* (*Carthamus tinctorius*), while he also adulterated *Musk* and the *Syrups*. The court to which such matters pertained, "Avogaria di comun," caused his preparations to be thrown over the Rialto Bridge into the canal, the *musk* was burned, the offender banished from his profession, placed in prison, and a fine of four hundred gold ducats ïmposed upon him—fortunately, *in contumaciam*.[1]

14. The proper study of drugs received an impulse from the chief emporium of the drug market of the middle ages.[2] The Signoria of Venice founded in the year 1533, in her University at Padua, a chair of pharmacognosy, "Lecturam Simplicium," which she filled by the appointment of the physician, Francesco Buonafede. The University of Bologna, as early as the year 1534, also received such a "Spositore o lettore dei Semplici" in the person of Luca Ghini, who, in the year 1544, removed to the University of Pisa. Buonafede, "primus Simplicium explicator," the first professor of pharmacognosy, distinguished himself by the foundation of the first botanical garden, which, in accordance with his proposal, in the year 1545, was effected by a resolution of the Venetian senate. The land required therefor was given by the Benedictine Convent, Santa Giustina, with a proper appreciation of the usefulness of the purpose. The garden still remains in sight of the magnificent cathedral of Sant' Antonio, in Padua. The decree of the senate, in a very admirable and appropriate manner, gives prominence to the value of the garden for the study of medicinal plants, and carefully provides for the management of the same. From the possessions of the Republic on the islands of Candia and Cyprus were to be procured the valuable plants, as also minerals.

[1] Cecchetti, "La Medicina in Venezia nel 1,300," Archivio Veneto, xxv. (1883), 376. *Ibid.*, xxviii. (1884), 29, it is also stated that, in the year 1320, on the Rialto Bridge, inspectors of goods from the Levant were appointed, "Signori sopra le merci del Levante."

[2] A description of this drug market is still wanting. As mentioned by Flückiger, *loc. cit.*, p. 634, there existed in Venice, in the year 1506, a chamber of commerce, "Cinque savii alla mercanzia," whose acts are still preserved in the great Venetian central archives.

As is evident from Gesner's "Horti Germaniæ" (see p. 32), he stood in active intercourse with apothecaries and other friends who were engaged in the cultivation of officinal plants. The opportunity was thus afforded of mentioning *cinnamon, cloves, worm-seed, colocynth,* and *fenugreek,* the mother plants of which were cultivated in a garden at Venice, in the quarter S. Gervasio.[1] But the first public garden for scientific purposes is that at Padua.

Buonafede also arranged in the botanical garden at Padua the first collection of drugs, "Spezieria," for purposes of instruction, in which the dried, crude products of the Levant were preserved, in order to be employed as standards for the discrimination of pure and adulterated articles. The garden and the collection are mentioned as two superabundant sources from which the most substantial knowledge may be obtained regarding those substances which contribute to the happiness of mankind. Notwithstanding these efforts, Buonafede, at the age of 76 (in the year 1549) lost his position, and, having become blind, died in penury on the 15th of February, 1558. The professorial chair, "lettura de' semplici," was taken in the year 1551 by Gabriel Falloppio.[2]

The University garden at Padua was followed in the year 1547 by that of Pisa, in 1567 by Bologna, in 1577 by Leyden, and in 1593 by Montpellier. In Germany the first botanical University garden was brought into existence in 1593 through the efforts of the medical faculty of Heidelberg; in the years 1624 and 1625 Ludwig Jungermann[3] established the gardens of Giessen and Altdorf (near Nuremberg). It was first in 1628 that Paris also received such an one. In the year 1658 there was a medical garden

[1] Flückiger, *loc. cit.,* p. 781. As early as 1330 the subject of a garden in Venice was agitated, which the celebrated physician Maestro Gualtieri wished to establish " pro herbis necessariis arti sue." Cecchetti, Archivio Veneto, **xxv.** (1883), 375.

[2] R. de Visiani, "Origine ed anzianità dell' orto botanico di Padova." Venezia, 1839, 43 pages ; and : "Della vita e degli scritti di Francesco Buonafede." Padova, 1845, pp. 24.

[3] Reess, "Address of the Prorector." Erlangen, 1884, 19.

in Westminster (London), which is presumed to have belonged to the "Society of Apothecaries." Prior to 1674, as it appears, the garden of this corporation was removed to Chelsea, where it still continues to exist.[1]

15. In the second half of the sixteenth and in the course of the seventeenth and eighteenth centuries, pharmacognosy was practically applied in the unnecessarily numerous pharmacopœias of different cities and countries of Europe, the largest number of which probably appeared in Germany. In this country, pharmacy had attained a high degree of development, as is shown by the official price-lists and various other ordinances. The first official pharmacopœia, however, entitled "Ricettario fiorentino," was published by the city of Florence in 1498. The number of drugs brought together in these pharmacopœias, dating from the sixteenth to the eighteenth century, was so considerable, that but little mention was made of new additions. Beside the most important of all, the *Cinchona-barks* (introduced in Spain about 1640, in England in 1655, and in Germany about the year 1669), there are to be mentioned, by way of example, *Ipecacuanha* (about 1682), *Catechu* (in 1640), and *Senega* (1735). Efforts now began to be made in the direction of the chemical investigation of drugs, which at the end of the seventeenth and beginning of the eighteenth century received the attention of numerous physicians and chemists; in Germany, *e. g.*, by Friedrich Hoffmann in Halle (1660 to 1742) and the distinguished apothecary to the Court, Caspar Neumann of Berlin (1683 to 1737), in Paris by the previously mentioned Geoffroy and Lémery (see p. 5), and in England by Robert Boyle (1627 to 1691).

Among the numerous and important discoveries of Scheele (1742 to 1786) there are but few which relate to drugs, although in this connection the discovery and examination of the acids most frequently occurring in plants was of the greatest importance. Scheele, in 1769, discovered tartaric acid, and in 1776 recognized the distribution of oxalic acid, which had already been observed by Savary in 1773; in 1784 he first prepared citric acid in a pure state, and in 1785 separated malic acid. Hydrocyanic acid was likewise discovered by Scheele in the year 1783, although it was not allotted to this celebrated investigator to also prepare this substance from plants.

Johann Bartholomæus Trommsdorff, who rendered such valuable services to pharmacy (from 1790 to 1837), and in 1795 established a Pharmaceutical School (Institute) at Erfurt, devoted a portion of his varied labors to the chemical investigation of vegetable medicinal substances. Notwithstanding these efforts, in his "Handbuch der pharmaceutischen Waarenkunde" (Erfurt, 1799, pp. 624; third edition, 1822), no essential advancement of pharmacognosy can be recognized.

The greatest progress in this department was inaugurated in the year 1816 by the apothecary Sertürner, through his discovery of the first alkaloid, morphine. Since this brilliant discovery, chemical research of the nineteenth century continues to give expression to the most important facts; and, for the reason previously intimated (p. 1), it more rarely occurs at the present time that a plant is able to attain permanent and prominent importance among valued medicinal agents. As accessions of the last eight decades are to be mentioned: *Carrageen, Cortex adstringens brasiliensis, Cort. Granati, Cort. Monesiæ, Flores Koso, Folia Coca, Gallæ chinenses, Guaraná, Gutta Percha, Helminthochorton, Herba Lobeliæ, Herba Matico, Kamala, Laminaria, Lupulin, Pengawar Djambi, Rad. Ratanhiæ, Rad. Scammoniæ, Secale cornutum, Semen Calabar, Sem. Colchici* and *Tuber Aconiti.* The very unequal, and in part very questionable importance of these drugs is at once apparent. As examples of drugs which have appeared more recently, and have directly received the condemnation of science, may be mentioned: *Lignum Anacahuite* (from *Cordia Boissieri* D. C., a Mexican shrub), *Cortex Condurango,* the leaves of *Sarrazinia purpurea,* of *Peumus Boldo,* of *Eucalyptus globulus* and of *Grindelia robusta,* furthermore *Cortex Coto, Cortex Quebracho* and *Radix Gelsemii.* On the contrary, as permanent and important acquisitions since the year 1873, may be considered:

Jaborandi leaves, from the Brazilian Rutacea *Pilocarpus pen-natifolius*, and the leaves of *Erythroxylon Coca*, from Peru and Bolivia.

The zealous endeavors of the eclectic school of medicine of North America to introduce new vegetable substances from their flora have also met with but little success.

After the discovery of strychnine and veratrine in the year 1818, of brucine in 1819, quinine, cinchonine and caffeine in 1820, coniine in 1827 (1831), aconitine and atropine in 1833, Liebig and Wöhler, in 1837, recognized in amygdalin the first representative of another numerous class of vegetable princi-ples, many of which possess energetic physiological properties. The chemical investigation of drugs, far superior to the anti-quated method of estimation, which was mostly based upon exter-nal characteristics, was thus correspondingly forced to the front.

16. In the mean time, not only the systematic investigation of the vegetable kingdom, but also its anatomical and physiological study became established more and more upon truly scientific principles. In consequence of this advance, the condition of pharmacognosy could not remain unchanged. Guibourt (1790–1867), in his lectures at the Paris school, and in his writings (see p. 5), had already, to some extent, taken a higher standpoint, and Pereira (1804–1853), perhaps to a still greater degree, in his large text-book (p. 5), in an introductory address in 1842, and, furthermore, in his lectures. Schleiden, in 1844, with perfect comprehension of the subject, demonstrated the importance of the microscopical examination of drugs, which, in 1847, was further elucidated in the most brilliant manner by an essay on the structure of *sarsaparilla*, accompanied by microscopic figures. When Schleiden's "Botanische Pharmakognosie," which was written in this spirit, made its appearance in 1857, the author no longer stood alone. Weddell, in 1845 and 1848, had become acquainted with the *Cinchonas* in Bolivia and Peru, and in 1849, in his handsome work on this subject, had also employed the microscope in an admirable manner for the purpose of distinguishing their barks. Schleiden extended this examination to all the *Cinchona barks* which at that time oc-

curred in commerce. To Otto Berg (1815–1866) is due the credit of having subjected a large proportion of the crude medicinal substances employed in Germany, at least all the more important ones, to microscopical study. In the figurative representation of the anatomical structure of organized vegetable substances he was, indeed, preceded by the no less distinguished Dutch botanist, Oudemans, 1854–1856. But, in 1865, Berg followed with the fifty plates, for the most part very true to nature, of his "Anatomischer Atlas zur pharmaceutischen Waarenkunde."

It is in this direction that pharmacognosy of the present day is being developed, on the one hand in practical connection with botanical science, and on the other supported by the constant progress in organic chemistry. The greater the number of active principles of the vegetable kingdom extracted by chemical processes, or even built up artificially, the greater the changes effected in the pharmaceutical importance of the respective drugs, for which, indeed, new interest may again be specially incited by a more searching botanical examination.

XII. **Pharmacognostical Systems.**—The larger number of medicinal substances, even by the most exhaustive treatment, will only appear of significance from some of the points of view that have just been mentioned, while in other respects they may present nothing at all worthy of note. It appears of less importance to treat of these substances in detail than to consider the order in which they are classified. They have been grouped in special pharmacognostical systems, in a more or less artificial manner, by considering either their organological importance, or, to a greater extent, their medicinal action and most prominent chemical constituents, or by utilizing at the same time all these principles of division. In opposition hereto, their arrangement in accordance with the natural families of plants, as followed by botanists, commends itself. The employment of a system founded upon this plan is appropriate, because a knowledge of the families of plants may be presupposed, thus scarcely leaving a doubt respecting the proper position of each drug, and because it does not admit of the separation of the parts or products which are furnished by one and the same plant. The advantages of such a classification are greater than the disadvantage, which may be supposed to arise from the fact that by this arrangement things are found in close association which are not connected either morphologically or medicinally. The "Pharmacographia," subsequently mentioned, as also the "Grundriss der Pharmacognosie" of one of the authors [1] (F.), presents the drugs of the vegetable kingdom arranged in accordance with the natural families of plants, while the "Pharmacognosie des Pflanzenreiches" (by F. A. Flückiger) bases its classification more upon the external characters of the drugs. An excellent example of chemical classification is presented by Falck's "Uebersicht der speciellen Drogenkunde," Berlin, 1883, p. 57.

[1] The "Grundriss" adheres to the well-planned system of the "Syllabus der Vorlesungen über specielle und medicinisch-pharmaceutische Botanik," by Eichler, third edition, Berlin, 1883, p. 54.

AIDS TO THE STUDY OF PHARMACOGNOSY.

The consideration of medicinal substances in their various aspects, as has been indicated, presupposes the possession of corresponding adjuncts or aids. These are primarily the requisite preliminary knowledge of botany, zoology and chemistry, as also experience in the use of the microscope. With this general knowledge, and the corresponding degree of skill, pharmacognostical description is everywhere intimately connected, as well as with the practice of pharmacy itself. The present work has not included the chemistry of plants in its plan, nor is it designed as a complete guide for microscopical study, and refers, therefore, in this respect to the literary aids enumerated under section II., *B*, page 47. In addition to the latter, an oral and practical introduction to the methods of microscopical examination is expressly recommended. Such opportunities are easily obtained, and, as in other departments of applied natural science, practical instruction proves here also to be of the greatest utility. Those who earnestly enter this field will soon become incited to zealous labor.

As scientific aids the following are to be considered:

I. COLLECTIONS.

A. Collections of Drugs.—The pharmacies themselves represent to a certain degree such collections, which are found more complete, in their adaptation to scientific purposes, in many of the higher educational institutions, and particularly in such as are directly subservient to the interests of pharmacy. The most instructive, and by far the most extensive collection of crude medicinal substances from the vegetable kingdom, however, is contained, in the Museum of Economic Botany, in the botanical garden at Kew, near London. Less extensive, since it is only in process of formation, is the botanical museum in Berlin. Collections which are devoted to pharmacy in its broadest sense are those of the Pharmaceutical Society of Great Britain, in their own building in London, the rapidly increasing collections of the General Austrian Pharmaceutical Association (Allgemeiner österreichischer Apotheker-Verein) in Vienna, and those of the School of Pharmacy (École de Pharmacie) in Paris.[1] Aside from the museum of the firm of Gehe & Co., in Dresden, which is worthy of consideration, the want of a large pharmaceutical centre in Germany is to be regretted. The Pharmaceutical Institutes connected with the Universities of Germany correspond but little to the ideal requirements which should be made of a pharmacognostical institution that is equipped in accordance with the times.

The great international expositions of modern times have presented, for the time, highly instructive and extensive collections of crude medicinal substances from different countries, of which reports have been published,[2] and in which, e. g., the medicinal agents even of the Asiatic nations are also thoroughly illustrated.

[1] This applies also, to a greater or less extent, to the museums and cabinets of some of the Schools of Pharmacy in the United States. (F. B. P.)

[2] For example, the excellent official reports of the Austrian experts who were delegated to these expositions. Compare further: Flückiger, Schweizerische Wochenschrift für Pharmacie, 1867, 325; also Archiv der Pharmacie, 214 (1879), 1–43 and 97–136. Paul, Holmes, and Passmore, "Universal International Exhibition," Paris, 1878; London, 1878, pp. 198. Schaer, "Botanischer Congress und Ausstellung pharmaceutisch wichtiger Pflanzenproducte zu Amsterdam," April, 1877. Archiv der Pharm., 212 (1878), 9–28. Wittmack, "Die Nutzpflanzen aller Zonen auf der Pariser Weltausstellung, 1878;" Berlin, 1879, pp. 112. See also the reports of Dr. Carl Mohr in the Pharm. Rundschau, New York, 1885, pp. 57–60, 77–83, 97–100, 126–131, 146–154, 165–170, 198–202, 227–230, entitled: "Mittheilungen über die medizinisch und technisch wichtigen Producte des Pflanzenreichs auf der Weltausstellung von New Orleans."

B. Collections of plants which are either themselves officinal, or which furnish the crude medicinal or technical substances,
or afford the material for preparations. Botanical gardens
present such plants in the living state with most admirable
selection and arrangement, for example, those of Kew (pages
12 and 44) and Edinburgh, the gardens of the École de
Pharmacie, and of the Faculté de Médecine in Paris, and
those of Berlin, Amsterdam, Vienna, and Palermo.[1] In the
garden of Mr. Thomas Hanbury at Mortola, near Mentone,
France, there are also cultivated many pharmaceutically important plants.[2]

Similar and very excellent results have been attained at the
garden of the University of Breslau through the judicious and
untiring efforts of Göppert.[3] The large Umbelliferæ of Asia,
which furnish *asafœtida, galbanum, sumbul* and *ammoniacum*,
are cultivated by Mr. Max Leichtlin in Baden-Baden.

Many plants of pharmaceutical importance are, as yet, to be
had only with difficulty or not at all, or are at least not readily
obtainable in the living state, so that all knowledge concerning
them must be derived from herbaria, or from descriptive and
illustrative representations. Such are presented in the greatest
degree of completeness by the previously mentioned institutes in
Kew, London and Paris; and the botanical museum (section, Her-

[1] In the United States, the gardens of Mr. Henry Shaw, at St. Louis,
Mo., are deserving of special mention, while the Congressional garden
and garden of the Department of Agriculture at Washington, the
botanical garden at Cambridge, near Boston, and the Arnold Arboretum
(for woody plants) at Brookline, Mass., likewise present something of
interest. There is also under contemplation the establishment of a large
botanical garden at Montreal, under the direction of the McGill University and the local horticultural society (F. B. P.).

[2] See Flückiger, "Osterferien in Ligurien;" Buchner's Repertorium
für Pharmacie, xxv. (1876), 449–505; also (Flückiger) "La Mortola, Der
Garten des Herrn Thomas Hanbury," Strassburg, 1886 (privately
printed), pp. 30.

[3] Compare his publications: "Unsere officinellen Pflanzen," Görlitz,
1883, pp. 12; and "Catalog der botanischen Museen der Universität
Breslau," Görlitz, 1884, pp. 54.

barium) in Berlin has also begun worthily to follow the example, though not precisely in a pharmaceutical direction.

C. Microscopical Preparations. Notwithstanding the importance of a knowledge of the inner structure of many medicinal substances, the preparation of objects for the microscope often requires such an expenditure of time as to interfere with its accomplishment. This is, however, so instructive (see p. 48) that it is only in cases of necessity, or for the purpose of supplementing individual work, that the purchase of microscopical preparations will be attended with substantial benefit. It must, nevertheless, be admitted that these preparations are now presented, especially in Germany, in an unusually attractive and complete form.

(see p. 48)

II. LITERARY AIDS.

The following list contains a selection of works which are adapted to quite extended requirements, but makes no claim to completeness.

A. Medico-pharmaceutical Botany. Of descriptive works, the following deserve particular mention:

Kosteletzky, "Allgemeine medicinisch-pharmaceutische Flora," 3 volumes, Prague, 1831–1834 (now sold by Hoff, Mannheim).

Geiger, Nees von Esenbeck and Dierbach, "Pharmaceutische Botanik," 3 volumes, Heidelberg, 1839–1843.

Bischoff, "Medicinisch-pharmaceutische Botanik," Erlangen, 1843; second edition, 1847.

Schleiden, "Handbuch der medicinisch-pharmaceutischen Botanik," Leipzig, 1852.

Rosenthal, "Synopsis plantarum diaphoricarum. Systematische Uebersicht der Heil-, Nutz- und Giftpflanzen aller Länder," Erlangen, Enke, 1872.

These works, which are excellent in their way, may, perhaps, occasionally still be consulted with profit; they are, however, more than replaced by the following:

Luerssen, "Medicinisch-pharmaceutische Botanik," also under the more appropriate title, "Handbuch der systematischen Botanik, mit besonderer Berücksichtigung der Arzneipflanzen," Vol. I., Cryptogams, pp. 657, 181 woodcuts, Leipzig, 1879. Vol. II., Phænogams, pp. 1,229, 231 woodcuts, Leipzig, 1882.

Luerssen, "Die Pflanzen der Pharmacopœa Germanica," Leipzig, 1883, pp. 664, 341 woodcuts. Chiefly compiled from the preceding.

Herz, "Synopsis der pharmaceutischen Botanik," Ellwangen, 1883.

Baillon, "Traité de Botanique médicale-phanérogamique," Paris, 1883–1884, pp. 1,499, with numerous woodcuts.

Targioni-Tozzetti, "Corso di botanica medico-farmaceutica," second edition, Firenze, 1847. Not seen by us. Furthermore:

Leunis, "Synopsis der drei Naturreiche." II. part, Botany. Third edition edited by A. B. Frank, Hanover, 1883.

Schimper, "Taschenbuch der medizinisch-pharmaceutischen Botanik und pflanzlichen Drogenkunde," pp. 208, Strassburg, 1886.

B. The following works give illustrations of officinal plants (accompanied by a descriptive text which is often unsatisfactory).

Nees von Esenbeck, "Plantæ medicinales," Düsseldorf, 1828–1833. 4 volumes. Folio, with 545 colored plates (12 inches wide and 19 inches long).

Hayne, "Getreue Darstellung und Beschreibung der in der Arzneikunde gebräuchlichen Gewächse, und solcher die mit ihnen verwechselt werden können," Berlin, 1805–1846. 14 volumes, quarto, with 648 colored plates.

Berg and Schmidt, "Darstellung und Beschreibung sämmtlicher in der Pharm. borussica aufgeführten officinellen Gewächse," Leipzig, 1854–1863. With 208 colored plates, quarto. Of excellent artistic execution.

Köhler's "Medicinalpflanzen," Gera-Untermhaus. Begun in 1883, and not yet completed.

H. Gross, "Die wichtigeren Handelspflanzen in Bild und Wort," Esslingen, 1880.

Bentley and Trimen, "Medicinal Plants," 4 volumes, London, 1875-1880. Small quarto, with 306 colored plates.

Artus, "Handatlas sämmtlicher medicinisch-pharmaceutischer Gewächse," etc. Sixth edition, edited by G. von Hayek. Jena, 1882.

Cassone, F., "Flora medico-farmaceutica," 6 volumes, with 600 colored plates. 8vo, Torino, 1847-1852. Not seen by us.

In America, the following illustrated works are in course of publication, which especially represent the medicinal plants in use in this country.

Lloyd, J. U. and C. J., "Drugs and Medicines of North America." A quarterly devoted to the historical and scientific discussion of the botany, pharmacy, chemistry and therapeutics of the medicinal plants of North America, their constituents, products and sophistications. Cincinnati, Ohio. Vol. I., of which the first number was issued in April, 1884, is now (March, 1886) complete, and comprises the natural order Ranunculaceæ.

Millspaugh, C. F., "American Medicinal Plants." An illustrated and descriptive guide to the American plants used as homœopathic remedies: their history, preparation, chemistry, and physiological effects (6 fasic.). Fasc. I., New York, 1884. Royal 8vo, 30 plates with descriptive text. (Four fascicles of this work have appeared up to the present date, July, 1886.) F. B. P.

C. Medico-pharmaceutical Zoology.

Brandt and Ratzeburg, "Medicinische Zoologie," Berlin, 1829-1833. 3 volumes, quarto, with 60 admirably executed copper plates.

Martiny, "Naturgeschichte der für die Heilkunde wichtigen Thiere," Giessen, 1854. 30 plates.

Moquin-Tandon, "Eléments de zoologie médicale," Paris, 1860; second edition, 1882, with numerous woodcuts.

D. Use of the Microscope.

Nägeli and Schwendener, "Das Mikroskop," Leipzig, 1867; second edition, 1877.

Dippel, "Das Mikroskop und seine Anwendung," 2 volumes,

Braunschweig, 1867–1872; second edition, 1883. By the same author, "Grundzüge der allgemeinen Mikroskopie," Braunschweig, 1885.

Hager, "Das Mikroskop und seine Anwendung." Seventh edition, with 316 woodcuts. Berlin, 1886.

Behrens, "Hilfsbuch zur Ausführung mikroskopischer Untersuchungen im botanischen Laboratorium," Braunschweig, 1883, pp. 398. With illustrations. American edition, entitled: "The Microscope in Botany," translated by Rev. A. B. Hervey, and R. H. Ward. Boston, 1885.

Stein, "Das Mikroskop und die mikrographische Technik," Halle, 1884. With illustrations.

Poulsen, "Botanische Mikrochemie," Cassel, 1881, pp. 83. American edition, entitled: "Botanical Micro-chemistry," translated by Professor William Trelease. Boston, 1884, pp. 118.

E. Chemistry.

Husemann and Hilger, "Die Pflanzenstoffe." 2 volumes, second edition, Berlin, 1882–1884. This book, which is very complete from a chemical standpoint, also elucidates the physiological action of the constituents of plants.

Dragendorff, "Die qualitative und quantitative Analyse von Pflanzen und Pflanzentheilen," Göttingen, 1882, pp. 285. English edition, entitled: "Plant Analysis, Qualitative and Quantitative," translated by Henry G. Greenish. London, 1884, pp. 280.

Ebermayer, "Physiologische Chemie der Pflanzen," Berlin, 1882.

F. Illustrations of Drugs.
Figurative representations of drugs in an ordinary sense are of less value than collections, perhaps with exception of the Cinchona barks, in which a comparison with illustrations may occasionally be desirable. Such colored illustrations of the Cinchona barks are presented in a very handsome form by the following:

Bergen (H. von), "Monographie der China," Hamburg, 1826. Quarto, with 7 plates.

Weddell, "Histoire naturelle des Quinquinas," Paris, 1849. Folio. The plates 28, 29, 30.

4

Delondre et Bouchardat, "Quinologie," Paris, 1854. Quarto. 23 plates.

A large number of drugs are illustrated, and described by valuable essays in the following:

Hanbury, "Science Papers," London, 1876.

With regard to some antiquated drugs reference may be made to Göbel and Kunze, "Pharmaceutische Waarenkunde," Eisenach, 1827–1834. 2 volumes, quarto, pp. 240 and 300. This work contains numerous colored plates, representing barks and roots.

G. Figurative representation of inner structure.

Oudemans, "Aanteekeningen op het systematisch- en pharmacognostisch-botanische gedeelte der Pharmacopœa Neerlandica," Rotterdam, 1854–1856, pp. 661, with 37 plates. Has long been out of print.

Berg, "Anatomischer Atlas zur pharm. Waarenkunde," Berlin, 1864, 50 plates. Quarto.[1]

Vogl, "Nahrungs- und Genussmittel aus dem Pflanzenreiche. Anleitung zum Erkennen der Nahrungsmittel, Genussmittel und Gewürze mit Hilfe des Mikroskops," Vienna, 1872, pp. 138, with 116 woodcuts.

Bell, "Die Analyse und Verfälschung der Nahrungsmittel." I. and II. Berlin, 1882 and 1884. The original English work, of which this is a translation, bears the title: "The Analysis and Adulteration of Foods."

Möller, "Mikroskopie der Nahrungs- und Genussmittel," Berlin, 1886, 8vo, pp. 384, with 308 woodcuts.

Schimper, "Anleitung zur mikroskopischen Untersuchung der Nahrungs- und Genussmittel," Jena, 1886, pp. 140, with 79 illustrations.

In the various commentaries to pharmacopœias, in the large work of Luerssen, mentioned on page 47, as also in monographs, there occur many illustrations pertaining to this subject. Thus, for example, in Rauter's "Memorial Publica-

[1] This excellent work will be further referred to by us under the designation of Berg's "Atlas."

tions of the Vienna Academy" ("Denkschriften der Wiener Akademie"), 31 (1872), 23, hair formations; Martinet, "Organes de sécrétion des végétaux" in *Annales des sciences naturelles*, xiv. (1872), 91, and many other publications mentioned in the chapter on anatomy. The structure of leaves is represented by Lemaire in his "Détermination histologique des feuilles médicinales," Paris, 1882; and that of barks in Möller's "Anatomie der Baumrinden," Berlin, 1882. The excellent illustrations accompanying Arthur Meyer's "Beiträge zur Kenntniss pharmaceutisch wichtiger Gewächse" in the *Archiv der Pharmacie*, Band 218, 219, 220, 221 (1881 to 1883), relate to the officinal Smilaceæ and Zingiberaceæ, to *Aconitum*, *Veratrum*, *Gentiana* and *Ipecacuanha*.

H. Special Pharmacognostical Text-books and Manuals.

Berg, "Pharmaceutische Waarenkunde." Fifth edition, Berlin, 1879, pp. 696. This first appeared in 1850 as the second volume of the second edition of his "Pharmaceutische Botanik" (the first edition of the latter was issued in 1845).

Flückiger, "Lehrbuch der Pharmacognosie des Pflanzenreiches," Berlin, 1867, pp. 758. Second edition, 1883, pp. 1,049.

Flückiger, "Grundriss der Pharmacognosie," Berlin, 1884, pp. 260.

Flückiger and Hanbury, "Pharmacographia: A History of the Principal Drugs met with in Great Britain and British India," London, 1874. Second edition, 1879, pp. 803.

Flückiger et Hanbury, "Histoire des Drogues d'origine végétale." Traduction de l'ouvrage anglais "Pharmacographia," par J. L. de Lanessan. 2 volumes, with woodcuts. Paris, 1878.

Fristedt, "Larobok i organisk Pharmacologi," Upsala, 1873.

Guibourt, "Histoire abrégée des Drogues simples," Paris, 1820. Sixth edition by G. Planchon: "Histoire naturelle des Drogues simples." 4 volumes, Paris, 1869–1870. With woodcuts.

Marmé, "Lehrbuch der Pharmacognosie des Pflanzen- und Thierreichs," Leipzig, 1885.

Oudemans, "Handleiding tot de Pharmacognosie van het

Planten- en Dierenrijk," Haarlem, 1865. Second edition, Amsterdam, 1880.

Planchon, "Traité pratique de la Détermination des Drogues simples d'origine végétale," 2 volumes, Paris, 1875. With illustrations.

Vogl, "Commentar zur österreichischen Pharmacopöe," Band I. "Arzneikörper aus den drei Naturreichen," Vienna, 1869. Third edition, 1880, pp. 516, with 164 woodcuts.

Wigand, "Lehrbuch der Pharmacognosie," Berlin, 1863, with 141 woodcuts. Third edition, 1879, pp. 447, with 181 woodcuts.

Wiggers, "Grundriss der Pharmacognosie," Göttingen, 1840, pp. 429. Fifth edition, "Handbuch der Pharmacognosie," 1864, pp. 800.

Wittstein, "Handwörterbuch der Pharmacognosie des Pflanzenreichs," Breslau, 1883, pp. 994.

The numerous drugs in use in India, especially those of the western portion of the peninsula, have been thoroughly considered by:

Dymock, "The Vegetable Materia Medica of Western India," Bombay (London), 1883–1884, pp. 786. (Compare the reviews in the *Archiv der Pharm.*, 222 (1884), 249, and *Pharmaceutische Zeitung*, Feb. 23d, 1884. Supplement to No. 16.)

All that has been accomplished in relation to the useful plants of India in the broadest sense, including the drugs, is compiled in the following large work in course of publication at Calcutta:

Watt, "A Dictionary of the Economic Products of India." The first volume of this work, comprising the letter A, was published in 1884. Large 8vo, pp. 353.

The drugs employed in the United States have been elaborated by:

Maisch, "A Manual of Organic Materia Medica," Philadelphia, 1882, pp. 459, with illustrations. Second edition, 1885, pp. 511, with 242 illustrations.

Lloyd (J. U. and C. G.), "Drugs and Medicines of North America." See page 48.

Of the resources of Brazil, an idea is given by:

Peckolt, "Catalog der pharmacognostischen, pharmaceutischen und chemischen Sammlung aus der Brasilianischen
Flora." *Zeitschrift des Allg. österreichischen Apotheker-
Vereines*, VI. (1868), 518. Extended pharmacognostical essays
by the same author are to be found in the periodical mentioned.

Representations and descriptions of drugs may be found,
although for the most part very deficient, in the various commentaries to the pharmacopœias, and in the manuals of the
practice of pharmacy.

J. An exhaustive work on the **History of Pharmacognosy,**
based upon an adequate study of sources of information, is still
wanting. In the circle of German literature the following are
worthy of consideration, however, in this respect:

Meyer, "Geschichte der Botanik," 4 volumes, Königsberg,
1854–1857. This work has, indeed, remained uncompleted,
since it extends only to the last quarter of the sixteenth century,
but nevertheless includes the periods which, in many respects,
are the most interesting for the history of pharmacognosy.

Jessen, "Botanik der Gegenwart und Vorzeit in culturhisto-
rischer Hinsicht," Leipzig, 1884. This work supplements the
last named, and contains, notwithstanding its brevity (495
pages), a large amount of substantial information, although
pharmacognosy does not properly fall within its sphere.

Hehn, "Kulturpflanzen und Hausthiere in ihrem Uebergang
aus Asien nach Griechenland und Italien, sowie in das übrige
Europa," fourth edition, 1882.

Sigismund, "Die Aromata in ihrer Bedeutung für Religion,
Sitten, Gebräuche, Handel und Geographie des Alterthums bis
zu den ersten Jahrhunderten unserer Zeitrechnung," Leipzig,
1884. (Reviewed in the supplement to the *Pharm. Zeitung*,
Bunzlau, May 31st, 1884, p. 377.)

A. de Candolle, "Origine des plantes cultivées, Bibliothèque
scientifique internationale," Paris, Germer Baillière & Co.,
1883.

K. **A Bibliography of Pharmacognosy** is also still wanting. The beginning of such a work is presented in Pereira's
"Elements of Materia Medica," II. (1857), 833–869. Some

references may furthermore be found in the previously cited (p. 51, under H), "Pharmacognosie," by Flückiger (especially pages 983–1022), and in the "Pharmacographia." A manuscript bibliography of pharmacy by Piper has been preserved by the Pharmaceutical Society of Great Britain in London since 1883.

L. That valuable aids to the study of pharmacognosy are contained in the various scientific journals and in monographs needs scarcely to be mentioned. In Germany, even previous to the year 1825, the *Archiv des Apotheker-Vereins im nördlichen Teutschland* inaugurated the stately series of volumes of the present *Archiv der Pharmacie*, while many other German professional journals have since ceased to exist. In Paris, the *Journal de Pharmacie et de Chimie* has likewise appeared regularly since that time. In England, Jacob Bell, who rendered such valuable service to English pharmacy (1810–1859), issued in July, 1841, the first number of the "Transactions of the Pharmaceutical Meetings," which now, under the title of *The Pharmaceutical Journal and Transactions*, continually presents a rich supply of pharmacognostical information, commensurate with the commercial status of England. The *American Journal of Pharmacy*, in Philadelphia, stands already in its fifty-eighth year of publication.

The "Annual Reports," finally, summarize all that is of value in current literature. Thus, in Germany, the "Jahresbericht der Pharmacie," which was begun in 1841 by Theodore Martius (1795–1863), and continued by Wiggers from 1844 to 1865. In 1866, Theodore Husemann, together with Wiggers, undertook its elaboration, and from 1867 to 1873 August Husemann was associated with them. In 1874, Dragendorff took the place of Theodore Husemann, and continued the "Jahresbericht" until 1878, when, in 1879, Marmé and Wulfsberg associated themselves with him. The "Bericht" of 1880 was edited by the latter, and the two following by Beckurts. A similar service has been rendered in England by the "Yearbook of Pharmacy," which has been issued since 1870, and much more punctually than the German "Bericht," by the British Pharmaceutical

Conference, and in America, since 1857, by the annual "Report on the Progress of Pharmacy," contained in the "Proceedings of the American Pharmaceutical Association," in which trade relations also find consideration.

' The *Botanisches Centralblatt* (Fischer, at Cassel), a journal of reference, which has appeared since 1880, under the editorship of Behrens and Uhlwôrm, and (since 1873) Just's *Botanischer Jahresbericht* (since 1883 under the editorship of Geyler and Kôhne), also consider the literature of pharmacognosy, even though only in a secondary manner.

MORPHOLOGY.[1]

The few officinal algæ, lichens and fungi, such as *Chondrus* (Irish-moss), *Cetraria* (Iceland-moss), *Secale cornutum* (Ergot), and the unofficinal *Fungus laricis* (Larch Agaric), afford examples of entire plants[2] which serve for medicinal purposes; the remaining drugs consist of parts of their mother-plants. Among the underground or, at least, half underground organs which fall within the province of pharmacy, the following are to be distinguished:

Roots (*Radices*). We confine this expression to the axes of endogenous formation, in which the capability of producing leaves, and mostly also chlorophyll, is wanting, but which possess a root-cap. This latter is a delicate, slightly extended tissue on the growing end (tip) of the root. The cells of the root-cap increase in number by division, but remain uniform; and farther back, within the cap, the point is reached where the formation of the different systems of tissue begins (see Anatomy, under formative tissue).

Roots are designated as main or *primary roots* when they represent the direct (underground) descending continuation of the base of the stem. Frequently the primary root becomes divided into root-branches, which, when they are very thin, are

[1] A thorough treatise on morphology, based upon modern views, is contained in Leunis' "Synopsis," Bd. I., 1882. Edited by B. Frank. Compare also Th. Liebe, "Die Elemente der Morphologie," Berlin, 1881.

[2] The sclerotium of *Claviceps purpurea*, which constitutes the *Secale cornutum*, is the passive condition of the fungus. *Fungus laricis* is the fruit-bearing portion of *Polyporus officinalis* Fr., and is thus, strictly considered, also only a portion of a plant, even though quantitatively by far the preponderating portion.

also called root-fibres. Examples of primary roots are afforded
by *Radix Pyrethri*, *Rad. Scammoniæ*, and *Rad. Taraxaci;*
excellent root-branches are possessed by *Rad. Ratanhiæ*, and
root-fibres by *Rad. Angelicæ*.

Secondary roots are such as originate laterally from the primary root or from parts of the stem, and which, upon the
whole, since the main root is often suppressed, occur much more
frequently than the primary root; in those cases, therefore,
where the distinction between primary and secondary roots is
not sharply characterized, as with *Rad. Angelicæ* and *Rad.
Levistici*, they may be designated simply as roots.

Occasionally, only the secondary roots are collected, but for
the most part the primary and secondary root together are officinal.

Above the tip of the root, mostly confined to small tracts, and
advancing with the growth in length, are found the root-hairs,
which individually (see Anatomy) become firmly attached in
their growth to the particles of earth, and convey the dissolved
nutritive substances of the soil to the root.

The function of the root is to fix the plant in the soil (hence
the central vascular bundle cylinder; see Anatomy) and to absorb the inorganic nutritive substances and water from the soil.
· The roots, as well as many rhizomes, are provided with a
bark which, in some cases, is removed by paring before the
drug is made use of (*Althæa, Glycyrrhiza, Iris*), and in others,
where it is rich in active constituents, it is allowed to remain
(*Calamus*). Barks separated from the root are also officinal,
thus, *e. g.*, *Granatum* (Pomegranate), the bark of the root of
Punica Granatum Lin. The root-bark of *Krameria* (Rhatany)
was also formerly in use.

Beside the true roots, there is still a series of underground
formations, which belong, however, morphologically to the
organs of the stem, even though they also fulfil physiologically
the function of roots. These are the runners or stolons, rhizomes and tubers.

The **Runners** (*Flagella*) and **Stolons** (*Stolones*) which
emanate from the root or from the base of the stem lie on the

surface of the ground or at a slight depth, and develop horizontally, often to a considerable length. They possess underleaves, at least in a rudimentary state, often contain chlorophyll, and inclose a clearly defined medulla. At some distance from the point of origin, the runners are able to develop roots and sprigs of foliage; if their connection with the mother plant is subsequently severed, they develop new individuals. *Radix Glycyrrhizæ* and *Rad. Saponariæ* consist to a large extent of runners.

The typical form of stolon is shown very nicely in the so-called *Rhizoma Caricis* (*Carex arenaria* Lin.). The sheath-like underleaves here surround the nodes, which are provided with root-bundles. *Rhizoma Graminis* (*Triticum repens* Lin.) may also be classed with the stolons.

The runners are botanically connected with the creeping or repent and decumbent stems.

Of the underground stem formations which bear under-leaves there are furthermore to be mentioned: the **Rootstocks** or **Rhizomes** (*Rhizomata*[1]). These are perennial stems or branches of vascular cryptogams and phænogams, growing half or entirely under the surface of the ground, which are provided with the rudiments of leaves, the remnants of leaf-sheaths or leaf-nodes, and send out roots (secondary roots). We also retain the name rhizome for those cases where the roots (secondary roots) are collected with it, or even when these preponderate, as with *Valeriana*. *Radix Sarsaparillæ* alone forms an exception; for although its rhizome occurs in part in commerce, it is nevertheless inadmissible for medicinal use.

The rhizome is provided at its apex with a bud, and annually develops new, ascending, herbaceous stems (leaves).

The peculiarity of the rootstock and the bulb, so definitely characterized by nature in opposition to the root, had already been rendered prominent by the Greek naturalist Theophrastos[2] (371 to 286 B.C.).

For the most part the rootstock alone is collected (*Iris,*

[1] 'Ρίζωμα root, stem.
[2] Compare Jessen, "Botanik der Gegenwart und Vorzeit," 1864, 26.

Galanga). When besides a central rootstock a secondary or lateral rootstock is present, both of them occur in commerce (*Zingiber*), although rarely as two separate commercial varieties (*Curcuma rotunda* and *longa,* the latter of which forms the secondary, the former the primary rootstock). Of *Zedoaria* only the primary rootstock is in use.

The roots of a rhizome (either alone or connected with the latter) form the sarsaparillas of commerce. [1]

Branched rootstocks (many-headed) are not of rare occurrence, and then possess many buds (*Galanga* [2]).

The rhizomes serve also as receptacles for reserve substances, and are nearly always densely filled with reserve nutritive substances (starch, etc.).

Still more pronounced receptacles for reserve materials are:

The Tubers (*Tubera*).—These are underground portions of the stem, or ramifications of the roots of phænogams (*Orchis*), which are so thickened that their diameter approximates their development in length or exceeds it. This growth in thickness stands in connection with the periodical accumulation of constructive materials, especially starch. They present a fleshy structure, which after drying is mealy and horn-like, not woody, as, *e. g., Tuber Aconiti, Tuber Jalapæ* and *Tuber Salep. Tuber Chinæ* deviates from the ordinary type of the tuber by its often very considerable length and strongly developed vascular bundles. [3]

It is, however, not only the stem which participates in the formation of underground organs, occasionally it is also leaves (under-leaves) which form these to a predominating extent, as, for example, in the case of the bulbs.

Bulbs (*Bulbi*).—These are fleshy, thickened under-leaves or parts of leaves, serving for the storage of starch and other substances serving as nourishment, which, similarly to buds, are severally inwrapped, scale-like, about a very short axis pro-

[1] Compare A. Meyer, Archiv der Pharm., 218 (1881), 280, with illustrations.
[2] A. Meyer, *loc. cit.* 425, with illustrations.
[3] Compare A. Meyer, Archiv der Pharm., 218 (1881), 272.

vided with roots, and stand above the surface of the soil or are half or entirely imbedded therein. The axis of the bulb is mostly flat, disk-like (Fig 1, *l*). The only officinal bulbs of the U. S. Pharmacopœia (1880) are *Scilla* and *Allium*.

By the accumulation of reserve substances, aid is given to the respective plant, corresponding to the climatic conditions, rendering it capable of attaining its full development in the shortest time and during the most favorable periods of the year.

A Corm (*Cormus*) is intermediate between the tuber and the bulb. It is a *bulbo-tuber* or *bulbodium tunicatum*, or a true tuber, which is provided with under-leaves and enveloped by them (*Colchicum*).

The thickened, and for the most part subterranean organs of

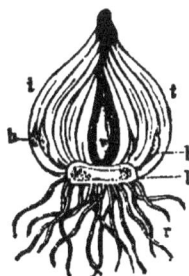

FIG. 1.—Median longitudinal section through a tunicated bulb. *l*, the disk-shaped. expanded axis; *v*, the bud; *t*, under-leaves; *r*, roots.

the stem, above described, lead us to the consideration of the proper organs of the stem, growing above ground.

Stem structures are formed either by the development of the stem portion of the plumule of the germinating plant (see below), or as lateral shoots on other organs of the stem, in the axils of leaves.

That portion of the stem which lies between the point at which the cotyledons are affixed and the point of attachment of the root is called the hypocotylous [1] member, caulicle or radicle. This remains mostly short (in the Hypogææ,[2] *i. e.*, those

[1] *'Υπό* under, and *κοτυληδων* (cotyledon) cavity.
[2] *'Υπό* under, and *γαία* earth.

plants whose cotyledons in germinating do not appear above the surface of the ground); it can, however, also become extended, and then elevate the cotyledons far above the ground (in the Epigææ,' for example, *Linum, Vicia faba*).

The form of a transverse section of the stem structure is cylindrical or angular. In the latter case the position of the leaves stands in direct relation to the number of the angles (*Labiatæ*, with a quadrangular stem and cross-like arrangement of the leaves). On cylindrical stems the leaves are arranged in spirals.

If the stems are developed in a leaf-like form, they are called phyllodia or phyllocladia² (*Ruscus*, Australian species of *Acacia Phyllocactus*).

In a pharmacognostical relation, the following organs of the stem come under consideration:

Stems or **Twigs** (*Stipites*). By these names are designated the weaker biennial and triennial, overground axes of dicotytedons, which are covered with an epidermis or cork, and contain chlorophyll. The only example properly considered here is presented by *Stipes Dulcamaræ*.

Woods (*Ligna*). By wood is understood the mostly "lignified" tissue (see below) of overground axes (or also of dicotyledonous roots), located inside of the cambium ring, which, with exception of the medullary rays, has attained considerable solidity through a material thickening of the cell walls. Only wood of many years' growth of the Gymnosperms, as also that of the Angiosperms, serves for medicinal purposes, with or without the bark, occasionally also either entirely or partially freed from the outermost, younger layers, the so-called sapwood or alburnum (see Anatomy).

Wood upon which the bark still remains, is frequently found in the two varieties of *Lignum Quassiæ* and in *Lignum Sassafras*, while *Lignum Juniperi* and *Lignum Guaiaci* are nearly always freed therefrom.

The numerous dye-woods, as, *e. g., Lignum Hæmatoxyli* (log-

¹ 'Επί upon or over, and γαία earth.
² Φύλλον leaf, and κλάδος shoot or stem.

wood) and *Lignum Fernambuci* (Brazil wood) arrive in commerce in the form of heartwood, freed from the colorless sapwood. In the case of *Lignum Guaiaci*, the valuable constituent, the resin, is also confined to the heartwood, although the dealers do not make a practice of removing the sapwood.

In a botanical relation, the herbaceous stems (*caulis*) are distinguished from the woody trunks (*trunci*), which, with the dicotyledons, as a rule, form branches (*rami*).

Barks (*Cortices*). The organs of the stem, as well as roots, are invested with a bark. The monocotyledons do not possess a true bark, since they show no secondary growth in thickness. The barks of dicotyledons—and such are officinal—consist in their young condition of a predominating parenchymatous tissue, *derma*, or primary bark, which is covered by the epidermis. Only barks of several years' growth, however, are brought into use, where the epidermis is replaced by cork. Where the latter is in a state of active growth, and represents a coherent layer, it is called periderm, outer bark, or *exophlœum*.[1]

Upon further development of the overground axes (or roots) which here come into consideration, the growth of the original bark is very soon increased through the activity of the cambium. Within the primary bark there is developed the secondary bark, inner bark, liber (bast), or *endophlœum*, the boundary of which is very frequently sharply defined from the primary bark. The bark then presents also a middle layer, the so-called middle bark or *mesophlœum*,[2] which, at a later period, often becomes very much diminished, constituting the remainder of the primary bark.

When, however, the formation of cork does not remain confined to the periphery, but forms also in the interior of the bark tissue, the middle bark, by means of cork bands, can become entirely separated and thrown off. This process, the formation of bork or rhytidoma,[3] can also extend to the inner bark. If, through peeling, the bork-scales themselves are also

[1] *'Eκ, ἐξ* out of, from, and *φλοῖον* bark.
[2] *Μέσος* middle.
[3] *'Ρυτίς, ῥυτίδος* fold, wrinkle; *δωμάω* I build.

removed, such a bark then consists at last almost exclusively of the inner bark, as, e. g., the *Cinchona calisaya* and *Coto bark.*

The inner bark is intersected by medullary rays (bark rays) which, upon a transverse and longitudinal section, are often plainly perceptible, without being magnified, as delicate lines.

The fracture of barks [that is, the appearance of a fractured surface], which is frequently a very useful characteristic, on the other hand, chiefly depends upon the fibres in the secondary bark (bast-bundles), upon the degree to which the elements of the same are thickened, and upon the form of combination of their prosenchymatous cells.

It is to very long, soft, and intertwining bast-fibres that *Cortex Mezerei* (Mezereum), as likewise the bark of *Radix Althææ*, owe their eminently fibrous character. The *Cinchona barks* are brittle, because their strongly thickened bast-tubes remain short and are usually isolated. The outer surface of *Ceylon cinnamon* admits of the ready recognition of the long, wave-like bast-bundles, which here and there intersect each other.

The cork (outer bark), is usually rejected, since it contains no active constituents. If it be thin, it often remains preserved on the bark, and can afford good means of distinction. Of the cambium, it is but seldom that remnants are preserved, since its cells possess very delicate walls.

Occasionally entire annual plants, or those of a few years' growth (without the roots) are employed. In pharmacognosy, these are then spoken of as herbs.

Herbs (*Herbæ*). These are the leafy shoots of phænogams which, beside the leaves and tenderer portions of the stem (the main axes are often removed) may possibly contain individual flowers and fruits, or the entire inflorescence and collection of fruits. There are no reasons for the exclusion of these structures which often accompany the leaves, and frequently it would be virtually impossible, as, e. g., in the case of *Herba Centaurii, Herba Meliloti, Herba Serpylli,* etc.

Not unfrequently, the form of ramification of the shoots is of diagnostic importance in the case of herbs. The different systems of ramification may, therefore, be considered here in a few words.

The formation of homologous axes is called ramification. An axis, together with its branchlets, is called a system of ramification. The systems of ramification appear in two principal forms.

1. THE DICHOTOMOUS SYSTEM (the axis ceases to increase in length at the apex):

(a) FORKED DICHOTOMY.[1] Directly beside the apex which extends no farther in growth, there arise two new axes which develop equally,. and can form the bases for further ramifications.

(b) SYMPODIUM[2] (in a more restricted sense). With each furcation an axis is developed, ultimately stronger than the other. The basal portions of the successive furcations form a pseudo-axis (sympodium) on which the more feebly developed axes are attached as lateral branches. Such a sympodium can consist either of the forked branches which are always of one side (bostryx), or of forked branches which are alternately of the right and left side (wickel or cinciunus).

2. THE MONOPODIAL SYSTEM[3] is produced by the main shoot continuing to grow, and forming lateral branches below the apex.

The monopodium can be either *racemose, i. e.*, when the main axis, even at later periods of growth, is always most strongly developed, or *cymose*, when the lateral axes overtop the main axis. In the latter case a pseudo-axis may be formed. Regarding this compare the subsequent remarks under inflorescence.

Exogenous[4] formations which arise laterally on the organs of the stem in acropetalous[5] succession, and have a different form from the stem which produces them, are called leaves.

The leaves arise densely crowded on the summit of the stem. If the parts of the stem located between two leaves become ex-

---

[1] Διχοτόμος cut in two (δίχα twofold) parts (τέμνω cut).
[2] Σύν entire, united, and πούς, ποδός foot.
[3] Μόνος single, and πούς foot.
[4] 'Εκ, ἐξ outside, and γένος formation.
[5] 'Ακρίς point, and *petere* to strive after.

tended, there are formed the *internodes*.[1] The parts of the stem bearing the leaves are thus called the nodes. These are occasionally swollen (*Polygonaceæ*, *Piperaceæ*, and *the grasses*). The place of attachment of the leaf is called the *point of insertion*.[2] According to the distribution of the leaves on the stem, there are distinguished the spirally arranged (*Chenopodium*), crossed or decussate[3] (*Labiatæ*), and whorled leaves (*Juniperus*).

The stalked leaves admit of the distinction of the leaf-stalk or

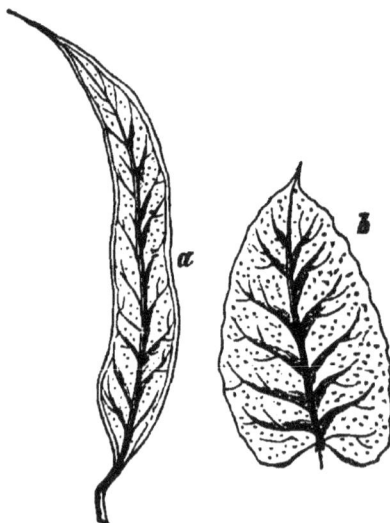

FIG. 2.—Leaves of *Eucalyptus globulus* Labill. (*Heterophylly*). *a*, sabre-shaped leaf of an older branch; *b*, heart-shaped leaf of a younger branch. The oil cells are indicated by dots. (Tschirch, in *Pharm. Zeitung*, 1881, No. 88.)

petiole (*petiolus*) and the leaf-blade or lamina;[4] on the boundary of both these parts there often appears a small membrane or ligule (*ligula*). The base of the leaf-stem is often developed as

[1] *Inter*, between, and *nodus*, node or joint.
[2] *Inserere*, to insert.
[3] *Decussare* (*decussio* = the number ten), to divide in form of an X (*i. e.*, crosswise).
[4] *Lamina*, a thin plate.

5

a broad sheath (*vagina*), and surrounds the stem in a tube-like
form. With the leaf-sheath are also to be classed the stipules
(*stipulæ*),[1] which are mostly of different, but occasionally of
the same form and color as the main leaves (*Asperula*), and occur at the base of the latter.

The form of leaves is very multifarious, although those which
are developed in a broad, flat form decidedly predominate.
If one and the same plant possesses differently formed leaves
at different places, it is called *heterophyllous*[2] (*Eucalyptus*,
Fig. 2). Since the leaves serve principally for the assimilation
of carbonic acid in the light, they present their broad, green
surfaces to the sun. This applies, however, only to the true

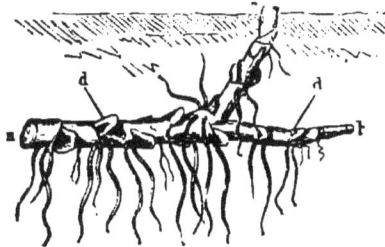

FIG. 3.—Rhizome of *Gratiola officinalis*, with roots. *a—b*, rhizome; at *b*, the terminal bud; *d*, under or rudimentary leaf; *c*, a shoot issuing from the axil of a rudimentary leaf, and likewise provided with rudimentary leaves.

FOLIAGE LEAVES (*Folia*), which are chiefly met with in the
region between the root and the flower. It is otherwise with the
UNDER-LEAVES or CATAPHYLLA (*squamæ*),[3] which are inserted
chiefly on the basal portion of the stem, and on the rhizomes,
stolons, etc., with a broad surface (Fig. 3). They are for the
most part brownish or pale in color, and to be regarded as
protecting envelopes for the tender, growing portions. The
fleshy bulb-scales are likewise rudimentary or under-leaves

[1] Properly halm or straw, on account of their dry, membranous con-
sistence. In German, *Nebenblätter*.

[2] Ἕτερος the other, and φύλλον leaf.

[3] *Squama*, scale.

(Fig. 1, *t*). Such appear, however, also as bud-scales, on the upper portion of the stem, enveloping and protecting the young leaf-buds; they have here the same function.

The germinating leaves or *cotyledons*[1] must also be regarded as rudimentary leaves. These possess, however, as we shall subsequently see, quite another function.

The third form of leaves are the HIGH LEAVES or HYPSO-PHYLLA (*bracteæ*).[2] These belong to the inflorescence, are likewise small and mostly delicate, and serve often as protecting integuments (the glumes of grasses).

The leaves, particularly foliage leaves, are often entirely (as with the stipules[3] of *Smilax*) or in part (many *Papilionaceæ*) converted into tendrils, although, on the other hand, all tendrils are not metamorphosed leaves (*Vitis, Ampelopsis, Passiflora*).

The most manifold forms and colors are assumed by the elements of the last leaf formation, **the flower.** While the leaves of the calyx (*sepals*) still possess a purely leaf-like nature and are mostly green, the leaves of the corolla (*petals*) appear colored and often of a particular form, the staminal leaves become converted into structures of quite another form (*stamens*), and the pistil-leaves closed together to form peculiar receptacles (*ovaries*). Nevertheless, the flower must be considered as the summit of a branch, encompassed by manifold leaf organs.

The summit of a branch or axis bearing the flowers (*receptaculum, hypanthium, torus*) is, as a rule, either conical or flat, but can also be enlarged in a broad or cushion-like form and become a disk[4] (*Rutaceæ*). If the disk lies within the circle of stamens, it is called an *intrastaminal* disk.[5]

There occur, however, also extrastaminal disks (*Æsculinæ*). The disk generally bears honey-glands or nectaries[6]; the latter,

[1] Κοτύλη cavity, pan.
[2] *Bractea*, a thin, small leaf of wood or metal. In German, *Hochblätter* or *Deckblätter*.
[3] In German, *Nebenblätter*.
[4] Δίσκος disk.
[5] *Intra*, within.
[6] Nectar, the drink of the gods, honey, a sweet substance.

however, also appearing as appendages of the perigon (*Ranunculus*), or as metamorphosed parts of the perigon itself (*spur*), or even as appendages of the stamens (*Cruciferæ, Lauraceæ*). The form of the nectaries is also exceedingly variable. In the case of *Euphorbia Cyparissias*, for example, they have the shape of a half-moon.

The flower-stalk or peduncle bears the flower (*flos*) which, in consequence of the bright colors of its corolla, attracts the insects for fertilization. This is the physiological significance of the color and fragance of the flower. Insects are especially attracted by the *labellum* of flowers of the Orchidaceæ (the posterior leaf of the inner perigon circle) and of the Zingiberaceæ (the metamorphosed lowest stamen of the outer staminal circle), as also by the *vexillum* of flowers of the Papilionaceæ (the uppermost petal).

When in a flower both the **calyx** and **corolla** are present, the leaves of the calyx or the sepals (*sepala*[1]) are mostly green, while the leaves of the corolla, or the petals (*petala*[2]), are of another color. When both cannot be distinguished, or one of them is wanting, the floral envelopes are spoken of as a *perigon*.[3] When the individual leaves are united or coalescent, the floral envelopes are termed sympetalous,[4] gamopetalous[5] (synsepalous, gamosepalous); when they are separate, choripetalous[6] (chorisepalous), or also spoken of in general as *gamophyllous* and *choriphyllous*.

According to the arrangement of the floral leaves in the bud, there are distinguished the following forms of æstivation :

Valvate (*Malva, Vitis*), Fig. 4 *a ;*

Imbricate (*Geranium, Veronica, Rosa*), Fig. 4 *b* and 6 ;

Convolute (*Gentiana, Phlox*), Fig. 5 ;

Plicate or plaited (*Campanula*), Fig. 7 ;

[1] *Separ*, divided (?)
[2] Πέταλον leaf.
[3] Περί around, and γόνος brood, seed.
[4] Σύν together, and πέταλον leaf.
[5] Γάμος marriage, and πέταλον leaf.
[6] Χωρίζω to divide (χωρίς separated), and πέταλον leaf.

and, according to the disposition of ordinary leaves in the bud, the conduplicate, plicate or plaited, involute, revolute, circinal or circinate, and corrugate or crumpled forms of vernation (*vernatio*). Compare also the diagrams in figures 11–15.

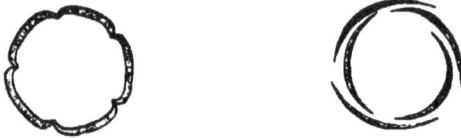

Fig. 4 *a*.—Valvate æstivation (*Vitis*).   Fig. 4 *b*.—Quincuncially imbricate
                                              ;æstivation (*Rosa*).

After flowering, the perigon, as a rule, falls off. In some cases, however, it takes part in the formation of the fruit. It remains either herbaceous (Chenopodiaceæ, Polygonaceæ), or becomes soft like a berry (*Morus*), or grows out in a hair-like form (*Eriophorum*). The calyx, by subsequent enlargement, often becomes the calyx of the fruit (*Hyoscyamus*, the Borraginaceæ, *Physalis*), and serves them for the protection of the lat-

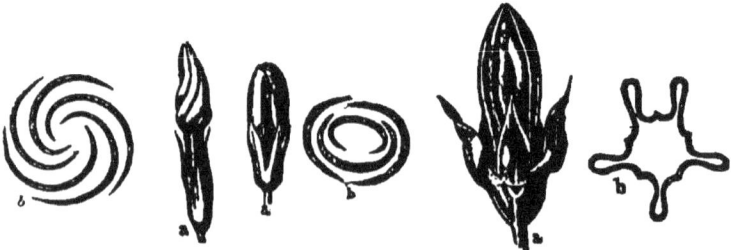

Fig. 5.   Fig. 6.   Fig. 7.

Fig. 5.—Convolute æstivation (*Phlox*) ; *a*, elevated view ; *b*, outline.
Fig. 6.—Imbricate æstivation (*Veronica*), *a* and *b* as in Fig. 5.
Fig. 7.—Plicate or plaited æstivation (*Campanula*), *a* and *b* as in Fig. 5.

ter. Occasionally, however, there appear in its place structural appendages which are comprehended under the name of *pappus*[1] (the Compositæ, *Valeriana*). The latter are either bristly or

[1] Πάππος crown of hairs on the fruits of the Dandelion and the Lettuce.

hair-like, and serve as organs of distribution, as well as of attachment of the fruits.

In plants of the Malvaceæ, there even appears, outside of
the true calyx, an involucre of high leaves or bracts, forming an
outer calyx (calyculus).   In *Fragaria*, the outer calyx is formed
from the stipules of the calyx leaves or sepals.

Occasionally the perigon is reduced to small scales (*lodiculæ*),
as in the case of many grasses (compare Fig. 11, *l*), certain species of *Aconitum*, and others.

The **andrœceum**[1] of the flower consists of the stamens.[2]
They form the male or fertilizing organs, and bear upon a long
stem (filament), which is occasionally branched or expanded in
a leaf-like form at the base, the receptacles of the pollen grains,
namely, the pollen-sacs in the anthers.[3]   These receptacles, which
are mostly two in number, constituting the so-called halves of the
anthers or *thecæ*,[4] are held together by the connective,[5] and, as a
rule, contain, in each, two pollen-sacs, lying one above the other.
There occur also so-called monothecous anthers (*Ricinus*, the
Malvaceæ).   It is rarely the case that one-half of the anther is
fertile, and the other not (*Salvia*).   The connective, which is
mostly short and but slightly developed, becomes occasionally
long and thread-like (*Salvia*), and then holds the anthers wide
apart.   In other cases (*Tilia*), it forms a wide bridge.   Rarely
are the anthers destitute of filaments, or sessile.

In order to discharge the pollen, the anthers open mostly by
a longitudinal line or chink (cleft, *rima*), more rarely by valves
(Lauraceæ).   If they dehisce toward the interior, they are called
introrse;[6] if, on the contrary, toward the exterior, *extrorse*.[7]   The
pollen is mostly granular, and readily becomes dust-like ; in the
Orchidaceæ and Asclepiadaceæ, it forms glutinous masses (*pol-*

[1] Ἀνήρ a man, and οἶκος house.
[2] *Stamen*, warp on a loom, thread.
[3] Ἀνθηρός blooming.
[4] Θήμη a receptacle.
[5] *Connectere*, to connect.
[6] *Introrsus* (*introversus*), toward the interior.
[7] *Extrorsus*, toward the exterior.

linaria, pollinia). The pollen grains, like to the spores of of the cryptogams, possess a variously marked outer coat (*exine*), provided with bristles or prickles ; by the development of the pollen tube, the *intine* projects in a sack-like form through the exine.

If the stamens have all coalesced to form one bundle, they are called *monadelphous*[1] (Geraniaceæ, Linaceæ, Oxalideæ); when coalesced to the number of two or several, polyadelphous[2] (Papilionaceæ). The stamens may, however, have coalesced with the perigon (*Symphytum*) or with the gynæceum, when they are called *gynandrous*[3] (Orchidaceæ, Aristolochiæ). In the latter case, the style, stigma, and anthers form the fructifying column (*gynostemium*).[4]

Occasionally the stamens develop leaf, horn, and pocket-like appendages which, in the case of *Asclepias* and *Vincetoxicum*, assume the form of an inner crown (*corona staminea*).

Sterile stamens are called *staminodiums* (Lauraceæ, *Linum*, Zingiberaceæ).

The **gynæceum**[5] (pistil[6]) of the flower, the female sexual organ of the plant, is formed of the fruit leaves or carpels,[7] which are mostly closed to form one vessel, the ovary, containing in its interior the seeds. If but one carpel takes part in the formation of the ovary, the latter is called *monomerous;*[8] the side traversed by the mid rib of the carpel is then the posterior side, the line of coalescence of the leaf-edges the *ventral suture*. Through the formation of false dissepiments or partitions, the monomerous ovary, which is usually unilocular, can become multilocular. When several carpels become united, the ovary is *polymerous.*[9]

[1] Μόνoς one, and ἀδελφός brother.
[2] Πολύς many, and ἀδελφός.
[3] Γυνή and ἀνήρ.
[4] Γυνή a woman, and στήμων stamen.
[5] Γυνή wife, γυναικεῖον female household.
[6] *Pistillum*, pestle.
[7] Καρπός fruit.
[8] Μόνος one, and μέρος part.
[9] Πολύς many, and μέρος part.

72   MORPHOLOGY.

If the ovary is *superior*, *i. e.*, if it occupies the uppermost part
of the flower, and the stamens and perigon are inserted below it,
the flower is called *hypogynous*[1] (Potentilleæ) (Fig. 8, *a*); if,
however, the andrœceum and perigon are elevated above the
ovary upon a cup-shaped, annular wall, by means of an axial
ring of the hypanthium located below the ovary, which latter
then remains at the base of the cup, the flower is called *perigynous*[2] (Roseæ) (Fig. 8, *b*); if the cup is now firmly closed at the
top, there is formed the *epigynous* flower[3] (Pomeæ) (Fig. 8, *c*).
In the two latter cases, the ovary is *inferior*.

The ovary is unilocular (Fig. 21) when the carpels are united
by their edges ; multilocular when the carpels are turned in-

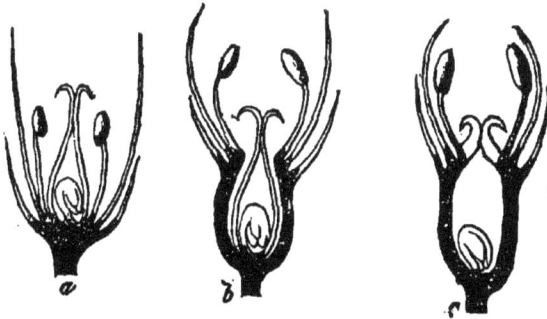

Fig. 8.—*a*, hypogynous ; *b*, perigynous ; *c*, epigynous flower, longitudinal section
(Prantl).

ward, and are in contact by their surfaces (Fig. 20). Through
the formation of false dissepiments, a polymerous ovary can also
become further divided (Labiatæ : 2 carpels and 4 compartments
or cells [nutlets or achenia], *Linum*), or a unilocular ovary can
become multilocular (Cruciferæ, *Papaver*). The false dissepiments, which, moreover, often produce but an incomplete divi-

[1] *Ὑπό* under, *γυνή* a woman (pistil).
[2] *Περί* about, *γυνή*
[3] *Ἐπί* upon, *γυνή*.

sion (*Papaver*), are for the most part extended growths of the placenta.

If a flower contains stamens and an ovary, it is called *hermaphrodite* [1] (☿ , as in most of the higher plants); when these are contained in separate flowers, they are called *diclinous* or *unisexual* [2] (Urticinæ). If diclinous flowers, the male (♂) as well as the female (♀), are found on one and the same plant, the term *monœcious* is applied [3] (Juglandeæ); if the male and female flowers occur on different plants, they are called *diœcious* [4] (Salicineæ, *Cannabis*, *Humulus*); when upon the same plant there are found unisexual as well as hermaphrodite flowers, the plant is called *polygamous* [5] (many Compositæ).

When several carpels occur, these are mostly coalescent in the ovary, as also in the style (syncarpous gynæceum, as in *Malva*). They can, however, be free from each other (apocarpous gynæceum, as in *Rubus*), and generally in all polycarpic flowers), or only coalescent in special places (partially apocarpous, as in *Asclepias Cornuti*).

The style, the upward extension of the carpels, bears the stigma. The stigma is either simple (Fig. 10) or branched (Fig. 9) (*Crocus*, Euphorbiaceæ), feathery, tufted (Gramineæ), often capitate, or even extended in a disk-like form, as in *Asclepias* and *Vincetoxicum*. [6] It is, as a rule, provided with papillæ (Fig. 9 in 3), and secretes a glutinous liquid. If the pollen grains of the anthers fall upon the stigma, they develop long tubes (pollen tubes, Fig. 10, *u*), which penetrate downwards in the conducting tissue of the style (*g*), as far as the ovules, and here effect the fertilization. The fertilized ovules then develop to form seeds.

Fertilization with other plants of the same species affords,

[1] Ἑρμαφρόδιτος bisexual.
[2] Δίς double, and κλίνη bed.
[3] Μόνος one, and οἶκος household.
[4] Δίς double, two, and οἶκος household.
[5] Πολύς many, and γάμος marriage.
[6] In the last-mentioned plants, the stigma also bears clamp-like bodies, to which the pollinia are attached.

even in hermaphrodite flowers, a greater supply of seed. Fertilization with the pollen of plants of other species produces, when successful, so-called hybrids. The transmission of the pollen takes place either through the agency of the wind (*anemophilous* plants), or more frequently, by insects (*entomophilous* or *zoidiophilous* [1] plants).

If a flower is projected on a plane, *i. e.*, if represented in such a manner that, when seen from above, only the points of inser-

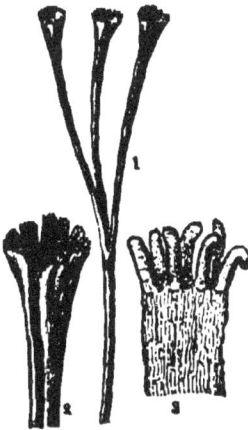 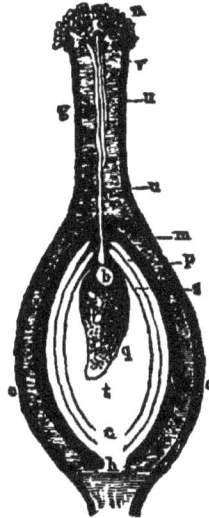

Fig. 9. Fig. 10.

Fig. 9.—*Crocus sativus.* 1, the three-limbed stigma ; 2, summit of a limb of the stigma, more strongly magnified ; 3, a section of the edge of the stigma, with the papillæ (Hager).

Fig. 10.—Schematic figure for the elucidation of the process of fertilization. *n* stigma with the papillæ (*u*), and three pollen grains already provided with protruding pollen tubes (*n*). Of the latter, one has already penetrated through the micropyle (*m*), and fertilizes the embryonal vesicle (*b*) contained in the embryo-sac (*q*). *p* outer, *s* inner integument ; *c* inner, *h* outer funiculus ; *t*, nucleus ; at a later period, occasionally passing into perisperm (the endosperm originates from the embryo-sac) ; *o*, pericarp.

tion of the separate parts of the flower are delineated by corresponding outline figures, a diagram[2] of the flower is obtained.

[1] Ἄνεμος wind, ζῷον animal, and φίλος friendly, loving.
[2] From διά through, and γράφειν to write.

In the diagram the stamens are represented by small circles, the staminodia or abortive stamens by crosses (Fig. 11), and the perigon leaves by segments of circles (Figs. 12 and 13). The abortive,[1] *i. e.*, the defective or suppressed parts of the flower, are designated by dotted lines.

Ordinarily, the individual parts of the flower are arranged in circles (cycles),[2] the members of which alternate with each other (Figs. 12 and 13), so, indeed, that each member of a successive circle, with regard to economy of space, always lies between two members of the preceding circle. In a typical dicotyledonous flower (Fig. 12), for example, the five corolla leaves lie between

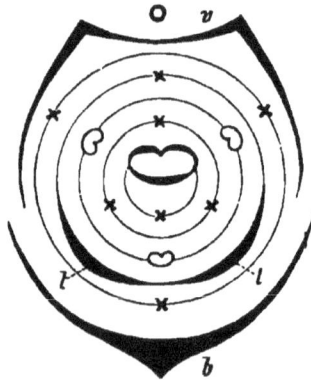

FIG. 11.

Diagram of a flower of the Gramineæ. *b*, palea inferior; *v*, palea superior ; *l*, lodiculæ or perigon leaves developed as small scales (Tschirch).

the five calyx leaves ; the first circle of stamens lies between the corolla leaves—therefore in front of the calyx leaves or sepals, the second circle in front of the corolla leaves or petals, etc. Such a flower is called *diplostemonous*,[3] as in most of the phænogams. If, however, the first circle of stamens lies above the corolla leaves (thus epipetalous), and the second above the calyx

[1] *Abortus*, miscarriage.
[2] Κύκλος circle.
[3] Διπλόος double, στῆμων stamen, thread.

leaves (thus episepalous), the flower is called *obdiplostemonous* [1] (Gruinales, Crassulaceæ, Saxifragaceæ).

According to the number of circles of the parts of the flower, the latter is spoken of as tri-, tetra-, or pentacyclous ;[2] and according to the number of members of the circle, as tri-, tetra-, or pentamerous [3] circles. The typical dicotyledonous flower is pentacyclous-pentamerous (Fig. 12), the typical monocotyledonous flower (Fig. 13) pentacyclous-trimerous.

In addition to a drawing (projection of the flower upon a plane), or by means of a diagram, the structure of the flower

FIG. 12.   FIG. 13.

FIG. 12.—Typical diagram of a dicotyledonous flower.
FIG. 13.—Typical diagram of a monocotyledonous flower (Liliaceæ) (Tschirch).

can also be expressed by formulas,[4] as *e. g.*, the typical dicotyledonous flower :

K (calyx) 5
C (corolla) 5
A (andrœceum) 5 + 5
G (gynæceum) (5)

which indicates that the andrœceum consists of 2 pentamerous circles, and the gynæceum of a superior ovary [(5)], formed of 5 coalesced carpels (Fig. 12). If a circle is deficient, it is de-

[1] *Ob*, opposite.
[2] *Τρίς, τέτρα, πέντε* (three, four, five), *κύκλος* circle.
[3] *Τρίς, τέτρα, πέντε* (three, four, five), *μέρος* part.
[4] See under Andrœceum in Eichler's " Syllabus."

signated by the mark 0, for example, in the flower of the Grasses (Fig. 11):

K 0

C 2

A 3+0

G (2)

or, since there is usually no differentiation of calyx and corolla in the monocotyledons [1] :   P (perigon) 0+2, A 3+0, etc.

If the ovary is inferior, it is expressed by a stroke above, for example, G $\overline{(3)}$.

 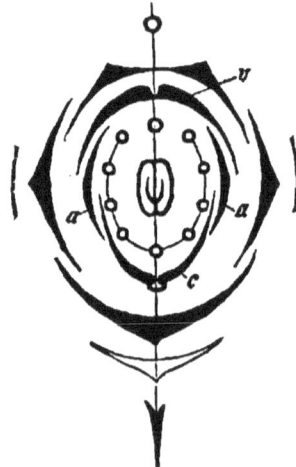

FIG. 14.   FIG. 15.

FIG. 14.—Diagram of a cruciferous flower, with diagonal corolla.

FIG. 15.—Diagram of a papilionaceous flower (descending æstivation in the corolla, ascending in the calyx). v, vexillum or standard ; $a\ a$, alæ or wings ; c, carina or keel (Tschirch).

Coalescence is indicated by parenthetical marks, as (2) = two coalesced leaves ; *deduplication* or *chorisis*,[2] French *dédoublement*,[3] *i. e.*, division of an organ into two or several parts by

[1] Μόνος alone, single, and κοτύλη.

[2] Χωρίζω to separate.

[3] *Dédoublement*, duplication.

an added exponent, which denotes the number of the parts, for example, 2^2.

If a flower can be divided into two equal halves (reflex images) by several dissecting lines drawn through its central point, it is called regular or *actinomorphous* [1] ($\oplus$) (Figs. 12 and 13). Such flowers, however, which can only be symmetrically divided by one section, are called monosymmetrical or *zygomorphous* [2] ($\wedge$) (Figs. 11, 14, 15).

If the only possible symmetrical section lies in the median plane (|), the flower is called median-zygomorphous ($\uparrow$), as in the Labiatæ and Papilionaceæ, otherwise oblique-zygomorphous ($\nearrow$), as in *Hyoscyamus*, or transverse-zygomorphous ($\rightarrow$), as in *Fumaria*. If flowers can in no manner be symmetrically divided, they are called *asymmetrical*,[3] as in the Zingiberaceæ.

The place where the bract accompanying the flower is located is represented in the diagram either below or in front. The first leaf of the flower is then located for the most part opposite the bract, either above or in the rear.

Through subsequent inversion (resupination), the position of the flower to the bract is occasionally reversed. Thus the labellum of an orchideous flower in its natural disposition is located in the rear and above, and only subsequently, through reversion of the ovary or of the peduncle, is brought forward and below.

The individual flowers often combine to form the so-called *inflorescence*. According to the form of ramification, the following are distinguished :

I. RACEMOSE INFLORESCENCE.—The main shoot (axis, rachis) is not over-topped by any of the lateral shoots produced thereon, in an acropetalous manner.

Elongated axis.
{
1. Spike (spica). The flowers sessile on the axis (Fig. 16 *a*) (*Carex*), to which belongs also : the catkin (amentum), when pendulous and falling off as a whole (*Juglans*).
}

[1] Ἀκτίς ray, and μορφή form.
[2] Ζυγόν yoke, and μορφή form.
[3] *A* privativum, and σύμμετρος symmetrical.

Elongated axis.

  2. Spadix. The flowers sessile, axis fleshy, and inclosed by a sheath (spathe) (Aroideæ).

  3. Raceme (racemus). The individual flowers have long pedicels (Cruciferæ) (Fig. 16 *b*).

Shortened axis.

  4. Head or Capitulum. The flowers sessile on a flattened or head-shaped axis (Compositæ) (Fig. 16 *d* and *e*).

  5. Umbel (umbella). On the axis, which is abbreviated to 0, originate numerous flowers with stalks or pedicels of equal length (Umbelliferæ). The umbel is often compound, so that small umbels (umbellets) stand at the end of every ray. The umbels and umbellets are then mostly surrounded by a circle of bracts, called the involucre or involucel (involucrum, involucellum) (Fig. 16 *c*).

II. CYMOSE INFLORESCENCE.—The main axis is over-topped by one or several more strongly developed lateral axes.

Without Pseud-axis.

  1. Cyme (cyma [1]). Below the terminal flower originate numerous, mostly equally strong lateral shoots (*Euphorbiæ*) (Fig. 16 *g*).

  2. Dichasium.[2] Below the terminal flower originate two equally formed lateral shoots (false dichotomy) *Valerianella* (Fig. 16 *m*).

With Pseud-axis.

  3. Bostryx.[3] The over-topping lateral shoots of successive members arise on the same side of their main axis (Fig. 16 *l*).

  4. Cincinnus.[4] The over-topping lateral shoots arise alternately on opposite sides of their main axis (Asperifoliaceæ) (Fig. 16 *h* and *i*).

By the combination of several types of inflorescence with each other is formed a compound inflorescence. To this belong the corymb (*Sambucus*) and the anthela.

[1] Κύμα the young stalk of the cabbage.
[2] From δίς twofold, and χάσις cleft, separation.
[3] Βόστρυξ curl, tendril.
[4] A curl of hair.

Finally, there is also to be mentioned the particular form of

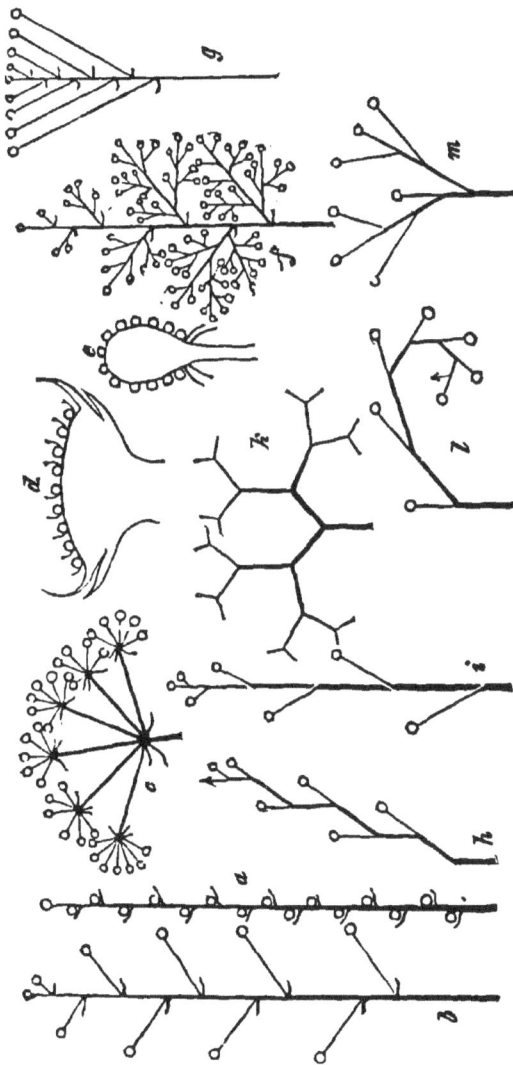

Fig. 16.—Schemes of inflorescences. *a*, spike ; *b*, raceme ; *c*, compound umbel ; *d* and *e*, heads ; *f*, compound raceme ; *g*, cyme ; *h*, and *i*, true dichotomy ; *k*, cincinnus ; *l*, bostryx ; *m*, false dichotomy.

inflorescence of the *Euphorbiæ,* which is designated as a cyathium. In this case, the cup-shaped hypanthium bears numerous male flowers (each with one stamen), and a long-stalked female flower.

Pharmacognostically, there belong to the flowers and forms of inflorescence, the developed, complete, officinal, separate flowers of the phænogams, and likewise the buds of individual flowers, for example, *Caryophylli*. Furthermore, undeveloped forms of inflorescence, *Flores Cinæ* (*Santonica*), as well as the expanded inflorescence, such as *Flores Arnicæ*, *Flores Chamomillæ*. *Flores Koso* (*Brayera*) consist of the inflorescence from which the petals have fallen. In the case of the composite flowers, the involucral scales are still present in the drug, perhaps with the sole exception of *Flores Arnicæ*. Finally, *Flores Rhœados*, *Flores Verbasci* and *Flores Rosæ* represent only the petals or corollas, and *Crocus* only the stigmas.

From the ovary with the ovules there is produced, after fertilization has taken place, the **fruit** containing the seed.

Fruits, aggregate or collective fruits, or parts of fruits of the angiosperms and gymnosperms are officinal with or without the seeds. For the rind (pericarp) of the Aurantieæ, which in the fresh state is juicy, the customary, though incorrect designation of *cortex* may be retained here, in order to avoid the introduction of a new term.

By the term "fruit" we mean here only the ovary which, as a result of fertilization, is in process of maturing, or has already become fully matured. Its outer wall, and the dissepiments and placenta, may thereby suffer the most manifold changes, by which other parts, including those not belonging to the flower, are also frequently affected, as, *e. g.*, in the case of the *fig*, *Fructus Juniperi*, the *apple* and *strawberry*, which are therefore to be designated as *pseudo-carps*.

In the *fig*, as likewise in the *strawberry* and the *apple*, the upper part of the axis participates in the formation of the (pseudo-) fruit (hence the name : hypanthodium [1]), which in all three cases assumes a fleshy character. In *Juniperus*, however, it is the three bracts of the flowers which become fleshy.

[1] From ὑπό under, and ανθος flower.

6

The wall of the ovary which becomes developed to form the
seed-vessel is called the *pericarp* (pericarpium [1]). On the latter,
from the exterior to the interior, there may often be distinguished three layers, differing in their structure or their color,
the *epicarp* (epicarpium [2]), *mesocarp* (mesocarpium [3]), and *endocarp* (endocarpium [4]).

The outer coating of the fruit often shows completely the
structure of the epidermis, *i. e.*, it is provided with a very strong
cuticle and with stomata ; often, however, it is formed to a
predominating extent of stone-cells (sclerenchyma). The variety
of the tissues and their contents is still greater in the middle
layer (mesocarp), which in many fruits consists of fleshy, juicy,
or even very loose tissue. When its cells contain an abundance
of juice and finally lose their coherence, it is designated as pulp,

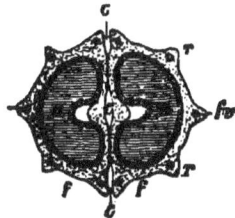

FIG. 17.—*Conium maculatum.* *a*, endosperm ; *c*, commissural surface ; *r*, costæ or,
ribs with the vascular bundles (*f v*); *f*, valleculæ or grooves (Hager).

as, *e. g.* in the legume of the *tamarind*. The inner layer of the
fruit (endocarp) originates from the epidermis of the cavity of
the ovary, and often develops into a hard stone-shell, as in the
case of the *almond*. It is, however, not always the case that a
discrimination can be made between the three layers of the
ripened seed-vessel, and their relative development is very
variable.

Of aggregations and forms of fruits, the following are distinguished :

[1] Περί about or around, καρπός fruit.
[3] 'Επί upon.
[4] Μέσος in the middle.
[4] Ενδον within.

The multiple fruit (syncarpium [1]), formed by the coalescence of several monomerous ovaries (*Star-anise, Rubus idæus*).

The mericarp (mericarpium [2]), double akene, is formed by the separation of the compartments of a multilocular ovary; the separate fruits thus produced are called schizocarps [3] (Umbelliferæ, Fig. 17).

With these may also be classed the jointed or articulated fruit (*Raphanus raphanistrum*), the mericarps of which are akenes.

The axis on which the two schizocarps of the Umbelliferæ are suspended is called the *carpophore*.[4] The main or primary

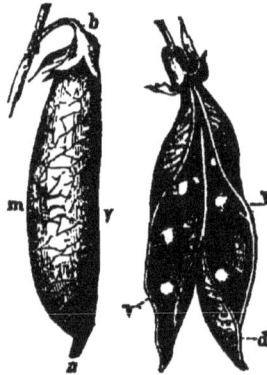

FIG. 18.—*Pisum sativum.* Legumes. a, tip; b, base; v, ventral suture; d, dorsal suture.

ribs of umbelliferous fruits are called *costæ* or *jugæ*;[5] the secondary ribs, *costæ secundariæ*; the grooves lying between them, *valleculæ*[6] (Fig. 184). In the latter are located, when present, the oil-tubes or *vittæ*.[7]

[1] Σύν together, καρπός fruit.
[2] Μέρος part, καρπός fruit.
[3] Σχίζω I split, καρπός fruit.
[4] Καρπός fruit, and φέρειν to bear.
[5] *Juga*, ridge.
[6] Diminutive of *vallis*, valley.
[7] *Vitta*, band.

There are furthermore distinguished :

I. Dry Fruits : Pericarp woody, coriaceous.

(a) Indehiscent fruits, not opening by valves or regular
    lines.

1. Nut (nux), having a hard pericarp (*Cannabis*).

2. Caryopsis[1] (or Grain) and Akene,[2] having a leather-
   like, membranaceous pericarp (Gramineæ, the cereals).

With these may also be classed the Samara, or so-called
winged-fruit, an akene which, through a subsequent growth of the pericarp, appears winged
(*Ulmus, Betula*).

3. Mericarps (see above).

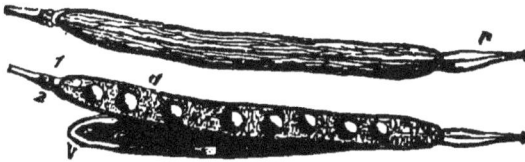

Fig. 19.—Silique of *Brassica oleracea*. 1, closed ; 2, opened, one valve removed ; *v*,
the other valve ; *d*, partition with the seed.

(b) Dehiscent fruits : opening by valves or regular lines,
    containing several seeds.

1. Follicle (folliculus[3]), formed of one carpel, and
   dehiscent by the ventral suture (*Illicium ani-
   satum*).

2. Legume (legumen), formed of one carpel, but also
   dehiscent by the dorsal suture (Fig. 18), often
   with false dissepiments (Leguminosæ).

3. Silique (siliqua), of two carpels. The latter separate from each other first at the base ; on the parti-

---

[1] *Κάρυον* nut, *ὄψις* appearance.

[2] *Ἀχαίνιον* (from α privative, *χαίνω* I open), a fruit which does not
open.

[3] Diminutive of *follis*, a sack or tube.

tion which remains attached, the seeds are located (Cruciferæ) (Fig. 19).

4. Capsule (capsula), formed of several carpels, dehiscent longitudinally from above (Fig. 20), more rarely from below, or ultimately opening by a lid (Pyxidium, Fig. 21 : as in *Hyoscyamus, Anagallis*), or by holes (pore-capsule, as in *Papaver somniferum*).

The dehiscence (dehiscentia) of the capsule is *septicidal*[1] (or through the dissepiments, in fruits opening by their sutures, Melanthieæ, Fig. 22 *a*), when in the case of a multilocular ovary the coalesced dissepiments become separated from each other

FIG. 20.   FIG. 21.

FIG. 20.—*Colchicum autumnale*, capsule dehiscent (septicidally) from above (Hager).
FIG. 21.—*Hyoscyamus niger*, capsule dehiscent by a lid. *a*, closed ; *b*, opened.

(*Colchicum, Sabadilla*); *loculicidal*[2] (in fruits dehiscing through the cells of the pericarp : Lilieæ), when each carpel becomes split in the middle (*Lilium, Scilla, Aloe*). If in the latter case the column of the dissepiments with the seed, separated from the wall of the capsule, remains standing in the middle, the dehiscence is called *septifragal*[3] (Fig. 22 *c*). In septifragal dehiscence, the opening may take place from below (*Geranium*), or from above (Balsamineæ, *Epilobium*).

II. Fleshy Fruits : Pericarp mostly fleshy.

[1] *Septum*, dissepiment, and *cædere*, to cut or to break.
[2] *Loculus* (diminutive of *locus*), compartment, and *cædere*.
[3] *Septum*, and *frangere*, to break.

1. The Stone Fruit or Drupe (drupa[1]), endocarp very hard, not dehiscent (*Amygdalus, Juglans*).

2. The Berry (bacca), endocarp and mesocarp fleshy, epicarp often hard (the grape, currant, date).

Occasionally the fruit is inclosed or surrounded by a body which is mostly cup-shaped, termed a *cupule* (cupula). In the case of the *oak*, this is formed from four coalesced bractlets. The tanning material known under the name of " valonia," consists of the cupules of the fruits of *Quercus Vallonea* Kotschy.

In the seed-vessel, formed of the carpels, are contained the *ovules* (ovula[2]). The latter consist of the *funiculus*[3] or podosperm, with which they are attached to the wall of the ovary, or to some special part of the ovary called the placenta,[4] which,

Fig. 22.—Forms of dehiscence of the capsule. *a*, septicidal ; *b*, loculicidal ; *c*, septifragal.

according to its position, is termed basal, central, or parietal ;[5] the *integuments*,[6] forming one or two coats (Fig. 10 *p s*), which in front do not close completely, but leave an opening (the micropyle,[7] Fig. 10 *m*); and the *kernel* or *nucleus* (Fig. 10 *t*), containing the embryo-sac (Fig. 10 *q*), in which the embryo[8] is formed (compare Figs. 23, 24, 25).

[1] *Drupus*, ripe and ready to fall.

[2] Diminutive of *ovum*, egg.

[3] Diminutive of *funis*, a rope or cord.

[4] *Placenta*, a cake, from a remote analogy with the placenta of the higher animals.

[5] *Paries*, wall.

[6] *Integumentum*, a covering or skin.

[7] Μικρός small, πύλη gate or entrance.

[8] Ἔμβρυον the unborn fetus in the womb.

These ovules, after having been fertilized, constitute the seed. That portion of the fruit of phanerogams which is formed from the ovule and contains the developed embryo is called the **seed**. Before the latter are utilized, they are, for the most part, completely freed from the seed-vessel. In the case of some seeds, the seed-shell or testa and the inner membrane are also removed.

The seed consists of the seed-coats or integuments and the embryo, and frequently, in addition thereto, of the albumen.[1] The former usually consists of an external, firm, occasionally very hard seed-shell or testa,[2] which is invested with a thin, but often very tough inner membrane ; this may readily be removed, especially after softening in water, as in the case of the *almond, coffee* and *Semen Ricini*, so that the kernel of the seed alone remains. *Semen Quercus* consists, in the commercial form, exclusively of the kernel, the two cotyledons without the membrane of the seed. With *Semen Myristicæ* or the nutmeg (as also with Cacao), on the contrary, the membrane penetrates into the kernel or nucleus, and in the case of the former, for example, cannot be separated in a connected form.

The testa is formed from the integuments of the ovule. For its first development the embryo requires a special supply of nutritive substances, which may be stored in the tissue of the embryo itself. In this case, a particular albumen is not present, the seed being thus destitute of albumen, or exalbuminous, as, *e. g., Semen Quercus*, the *almond* and *mustard* (and in general all the Cruciferæ).

If, however, there is developed, simultaneously with the embryo, a special tissue filled with reserve material, this is called albumen. If this tissue, with regard to its origin, belongs to the embryo-sac, as is usually the case, it is termed endosperm[3] (Umbelliferæ, Fig. 17 a); if, however, a portion of the nucleus (kernel of the ovule) has been converted into albumen, it is dis-

---

[1] *Albus*, white (the white of egg).
[2] *Testa*, a vessel, or also a shell.
[3] Ἔνδον inside, σπερμα seed.

tinguished as perisperm.[1] The seeds of the *cardamom* and of *pepper* represent simultaneously both forms of albumen, perisperm as well as endosperm. The chief contents of the albumen-cells belong nearly always to the class of protein substances, and are frequently developed, in part, in a crystalloid form. With these is usually associated fat, and not rarely also amylum, sugar, and mucilage. This abundance of contents, which, moreover, are very commonly deposited in thick-walled cells, mostly imparts to the tissue of the albumen a firm, hornlike character. In the German usage of the word, there is accordingly sometimes understood (rather ambiguously) under the expression "albumen," the entire tissue of the seed which contains the previously mentioned reserve substances, and sometimes, in a chemical sense, that class of nutritive substances which are also termed protein bodies.

The degree of development of the albumen is very variable. It is often much more extensive than the embryo, as in *Semen Myristicæ* (the nutmeg), *Semen Colchici*, and in *Nux vomica;* in other cases, it appears only as an insignificant appendage, as in *Semen Lini*, or perhaps also disappears at a later period, so that it is no longer observable in the ripe seed.

The embryo contains, in a more or less advanced state of development, the rudiments of the axis and leaf-organs, the former shortly attenuated in one direction as the radicle or caulicle (this is always directed toward the micropyle), and in the opposite direction often bearing the rudiments of the stem and leaf structures, or plumula ;[2] the latter may very plainly be seen, for instance, in the *almond*, and also in *Nux vomica.*

The leaf-organs, embryonal leaves, seed-lobes or *cotyledons*,[3] usually form the preponderating portion of the embryo, and occur, especially in many dicotyledons, already developed in a delicate, distinctly leaf-like form, as in *Nux vomica* and *Semen Ricini*. In seeds destitute of albumen, *e. g.*, in the *almond*, *bean*, *pea*, and *acorn*, on the contrary, the cotyledons are of a

---

[1] Περί around or about, and σπέρμα seed.
[2] Diminutive of *pluma*, feather.
[3] Κορύλη cavity, κορυληδών cavity of a bone, pan.

thick, fleshy character. With monocotyledonous plants, the embryo in the seed is usually less clearly developed ; in *Semen Colchici, Semen Sabadillæ*, and in the *cardamom*, the cotyledon is not yet really distinctly leaf-like, and quite as little in the *pepper* and *cubeb*. The tissue of the embryo throughout is built up of more delicate cells than that of the albumen, and this difference is also readily apparent without being magnified.

The cotyledons and the radicle are often bent in a character-istic manner, as is evident upon a longitudinal section through *Semen Stramonii*, while the fruits of the Umbelliferæ present an example of a tolerably straight embryo. A very remarkable folding is shown by the embryos of *Semen Fœnugræci* and *Semen Sinapis*, as in general with all Orthoplocæ, Spirolobeæ and Diplecolobeæ. Outside of our sphere of consideration, there occur remarkably complicated foldings in the cotyledons of the cotton-seed.

The seed is connected with the placenta by means of the funiculus ; the place where the latter enters the testa usually remains characterized by its color, a depression, or an elevated line, and is distinguished as the *hilum* (Fig. 10 *h*, Fig. 24 *h*). Less frequently the terminal point of the funiculus is also per-ceptible in the base of the seed ; if this is the case, it bears the name of inner hilum or *chalaza* [1] (Fig. 24 *ch*). This is readily recognizable, among other instances, in *Semen Ricini*.

The seed is straight, *atropous* or *orthotropous*,[2] when the apex of the ovule, the orifice (micropyle), lies opposite the hilum, whereby the funiculus remains short. It is thus, *e. g.*, with the Piperaceæ, where the seed forms the termination of the flower axis. More frequently, however, the ovule together with its coats, *i. e.*, the entire seed, is reversed, whereby its apex, the micropyle, is moved close beside the hilum. This form, with the funiculus running along the back, which is the most usual with the angiosperms, is designated as a reversed, anatropous [3] seed (Fig. 24). The ovule is here coalescent with the funiculus

---

[1] Χάλαζα hail, but also a sty on the eye-lid.
[2] From α privative (and ὁρϑός straight), and τρέπω turn, direct.
[3] Ἀνά against, τρέπω I turn.

whereby a suture, *rhaphe*,[1] is produced (Fig. 24 *r*), which is more or less apparent, *e. g.*, in *Semen Tiglii* and in the *cardamom*.

The reniform or kidney-shaped seeds, on the contrary, are mostly produced from so-called *campylotropous*[2] or curved ovules. With these, the nucleus as well as the integuments are curved (Fig. 25). They therefore also possess, as a rule, a curved embryo.

With relation to their attachment, the ovules are sometimes pendulous, sometimes erect or ascending, and sometimes horizontal. If in an anatropous (pendulous) ovule the funiculus lies toward the interior or the middle-line of the fruit, such an ovule is termed *epitropous*[3] (Umbelliferæ, Euphorbiaceæ). If the funiculus is directed toward the outer wall, it is called *apotropous* (*Vitis, Rhamnus, Cornus*).

FIG. 23.   FIG. 24.   FIG. 25.

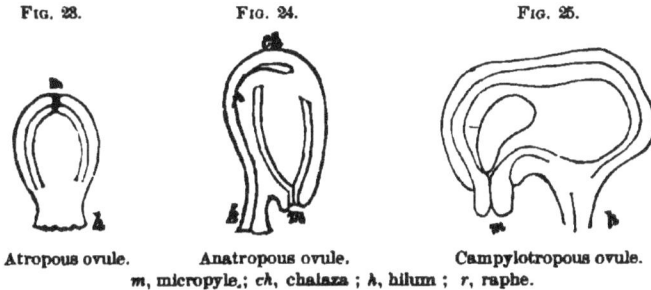

Atropous ovule.   Anatropous ovule.   Campylotropous ovule.
m, micropyle.; *ch*, chalaza ; *h*, hilum ; *r*, raphe.

Appendages of the Seed.—Many seeds are provided at the hilum with an indurated appendage (as in the *pea*), which, in the case of *Semen Ricini, Semen Chelidonii* (obdurator, caruncle or caruncula[4]) and *Semen Colchici*, still remains perceptible, even after drying, while in *Semen Tiglii*, on the contrary, it readily falls off.

A peculiar, compact, fleshy outgrowth is developed on the *nutmeg*, and is designated as the seed-covering or *arillus* (Fig.

[1] *'Ραφή* suture.
[2] *Καμπύλος* bent, and *τρέπω*.
[3] *'Επί* upon, and *τρέπω*.
[4] *Caro*, flesh.

26 *a r*). This arillus, which is known in commerce under the name of *mace*, represents the only structure of this character which is properly treated of here. A relatively still more developed arillus, but consisting only of a thin membrane, incloses the seed of the *cardamom*. The small, red, cup-shaped body of *Taxus* fruits is also to be regarded as an arillus.

The arillus always originates at the base of the seed, and is to be considered as an outgrowth of the funiculus.

The pappus, a form of appendage of fruits (page 69), is formed by a subsequent outgrowth of the calyx.

The germination of the seed proceeds in this way, that with a

Fig. 26.—Fruit of *Myristica fragrans*, longitudinal section. *ar*, arillus; *s*, seed (Hager).

simultaneous evacuation of the endosperm which may be present, the plumule and radicle break through the testa of the seed, the former developing to form the stem, and the latter to form the root. Thereby the cotyledons, which are sometimes fleshy and thick (as in the *Bean*), sometimes thin and leaf-like (*Ricinus*), are either elevated above the ground and become green (Epigæa), or remain in the ground until their evacuation and rejection (Hypogæa, compare page 69). Occasionally, the cotyledon still incloses for some time the young leaf-bud (*Maize*).

PLANT ANATOMY.[1]

In order to obtain a satisfactory knowledge of vegetable drugs, an accurate anatomical study of them is in most cases indispensable. This part of pharmacognosy is therefore based upon an acquaintance with the principles of plant anatomy. The following lines may serve for a preliminary acquaintance with this very extended department, more complete information being contained in the text-books of anatomy.[2] The beginner should, nevertheless, continually bear in mind that anatomical study, unless accompanied by work with the microscope,[3] must always remain poor and deficient in its results. It would, there-

[1] From ἀνά and τέμνω cut.

[2] De Bary, "Vergleichende Anatomie der Vegetationsorgane," Leipzig, 1877. The most comprehensive and fundamental work, which, with regard to the amplitude of its contents, can be compared with no other. An excellent English translation of this work bears the title : "Comparative Anatomy of the Vegetative Organs of the Phanerogams and Ferns;" by A. De Bary. Translated by F. O. Bower and D. H. Scott, 1884 (F. B. P.).—Sachs, "Lehrbuch der Botanik," iv., Leipzig, 1874 (at present only to be had through antiquarian book-sellers). An English edition of this work bears the title : "Text-book of Botany," by Julius Sachs. Translated by A. W. Bennett, assisted by W. T. T. Dyer, Oxford, 1875. Second edition, 1882. F. B. P.—Haberlandt, "Physiologische Pflanzenanatomie," Leipzig, 1884.—Weiss, "Anatomie der Pflanzen," Vienna, 1878.—Leunis, "Synopsis," newly edited by Frank. One volume, Hannover, 1882. For our purpose, there may also be mentioned : Hanausek, "Anatomische, physikalische und chemische Verhältnisse des Pflanzenreiches, mit besonderer Rücksicht auf Warenkunde und Technologie," Hölder, Vienna, 1882.

[3] In microscopic work, the following are very useful : E. Strasburger, "Das botanische Practicum," Jena, 1884 ; and, by the same author, "Das kleine botanische Practicum," Jena, 1884.

fore, really seem most proper to provide this chapter with an introduction to the construction and use of the microscope. There have recently appeared, however, so many publications relating precisely to this subject,[1] that we may omit further reference to it in this place.

The experience acquired at the preparation table is in the end the most important of all ; and as chemical analysis cannot be learned without a laboratory, so also the methods of microscopical investigation cannot be learned without a microscope and dissecting needles.

I. The Cell.

The elementary organs from which the body of the plant is constructed are the cells. Although it is not necessary for the formation of the idea of a cell that the same should be inclosed by a membrane (naked swarm-pores), nevertheless, by far most cells are provided with such.

Most plants (all the more highly organized ones) consist of numerous cells. Among the lower plants there are, however, many which are formed of but single cells, some of which assume the most manifold forms, branch abundantly (the mould fungus *Mucor Mucedo*), and, indeed, without being in any manner divided by lateral walls, imitate a stem, leaf, and root (*Caulerpa*). Of such one-celled plants, there are none which come under consideration in pharmacognosy in a restricted sense, although the yeast fungus (*Saccharomyces cerevisiæ*), on account of its fermentative action, and the various pathogenic fungi (*bacteria*), which claim an increased degree of interest in consequence of the recently observed relations between them and the most dangerous diseases, as well as the Diatomeæ, whose siliceous coats

[1] Behrens, "Hilfsbuch zur Ausführung mikroskopischer Untersuchungen," Braunschweig, 1883. The American edition of this work has been noticed on page 49, F. B. P.—Dippel, " Das Mikroskop," II., Braunschweig, 1883–84.—For the theoretical part, the following is very valuable : Nägeli and Schwendener, "Das Mikroskop," Leipzig, 1877.— Further : Hager, " Das Mikroskop und seine Anwendung," Berlin, 1879.—J. Vogel, "Das Mikroskop," Leipzig, 1885.

form the so-called infusorial earth, are of the greatest importance in practical life.

THE CELL-WALL AND CONTENTS OF THE CELL.

I. Contents of the Cell.

The cell consists of the cell-wall, and the contents of the cell.
The most essential constituent of the cell while exercising the
functions of life, is the **protoplasm** (plasma [1]). This represents a turbid, semi-liquid mass, which completely fills the inner
portion (lumen) of cells located at the growing point, or which
are otherwise in a state of active development. At a later
period, there appear in the protoplasm cavities or vacuoles.[2]
The latter, filled with colorless cell-sap, constantly continue to
enlarge as the cell becomes older, coalesce with each other, and
finally, while the protoplasm gradually contracts toward the
wall of the cell (protoplasm-sac, primordial utricle, cell-sac),
form a large, central cavity, filled with cell-sap (Fig. 27 p). If
the cell has ceased to grow, the protoplasm will also have disappeared, with the exception of a delicate film attached to the
membrane. The protoplasm takes the most active part in all
formative processes in the cell, and is the most important substance in the cell ; the formation of the cell-wall proceeds from
it, and to it the other constituents of the cell, for the most part,
owe their origin.

The cells increase in number only by division. A cell (mother-
cell) becomes divided by a septum (which is mostly median)
into two daughter-cells (Fig. 28). And it is considered as a law
that with the division of the cell (of the protoplasm) a division of
the nucleus (see below and Fig. 28) is also always associated. In
the case of reproductive cells, a rounding of their form is also
generally associated with their division.

The protoplasm, which always possesses a semi-liquid, and
(with the exception of the outermost and innermost layer, hyaloplasm) a granular character (microsomes), is a body of compli-

---

[1] Πρῶτον the first, and πλάσμα organization or form.
[2] Vacuum, empty.

cated composition, very rich in nitrogen,[1] which contains several substances belonging to the albumen group (protein substances), together with water and inorganic salts (phosphates and sulphates of the light metals). It is not devoid of structure, but possesses a fine organization.[2]

Substances capable of abstracting water (such as sugar and glycerin) contract the protoplasm, *i. e.*, in consequence of the

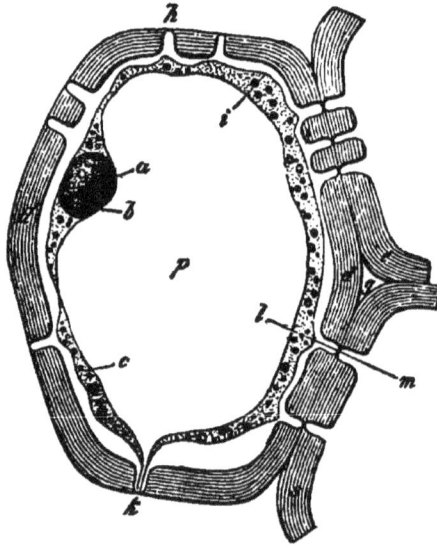

FIG. 27.—Transverse section through a medullary cell of *Toxodium distichum*. *a*, nucleus; *b*, nucleolus; *c*, protoplasm-sac contracted toward the wall (separated from the latter by reagents); *i*, primordial utricle (hyaloplasm); *p*, cell-sap; *l-m*, corresponding tips of adjacent cells; *d*, the cell-wall; *e-s*, the cell-walls of adjacent cells; *g*, intercellular space (Hartig).

elimination of water from the contents of the cell, the protoplasm-sac is drawn from the cell-wall. Protoplasm is colored

[1] Compare Reinke, " Studien über das Protoplasma," Berlin, 1881.

[2] Very many investigations have recently been published, relating to the structure of protoplasm (especially by Strasburger, Schmitz, Tangl, Formmann, and others).

yellowish-brown[1] by iodine,[2] rose-red by Millon's reagent, violet
by Trommer's reagent, and red by sugar and sulphuric acid.
Dead protoplasm abundantly absorbs coloring substances (especially fine with eosin). Imbedded in the protoplasm of young
cells (Fig. 27), or suspended by threads of protoplasm (Fig. 28,
b, c), there is found the *nucleus* (Fig. 27 *a*, and Fig. 28), which
mostly occurs singly. The nucleus possesses one or two *nucleoli*
(Fig. 27 *b*), and consists likewise, for the most part, of a proto-
plasm-like substance, in which, however, the *nuclein* is contained in a granular form. By treatment with coloring substances (hæmatoxylin, aniline-green, alum-carmine), the nuclei
are rendered more clearly visible.

Since the protoplasm has contracted in old cells to the mini-

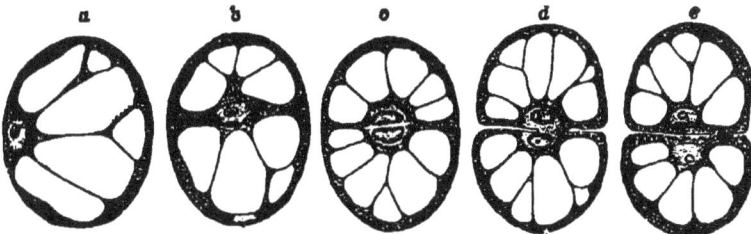

Fig. 28.—The process of cell division schematically represented in its individual phases
(Hartig).

mum and the nucleus has entirely disappeared, both of these play
but a subordinate part in pharmacognosy, which occupies itself
mostly with organs consisting of completely developed tissues,
notwithstanding the importance of the protoplasm in the economy of the plant itself, and in the estimation of the value of
herbs as fodder.

[1] All protein substances are colored yellow by iodine, thus protoplasm,
gluten, protein crystalloids (for the latter, iodine in glycerin is employed), the fundamental portion of the chlorophyll granules, etc.

[2] For these micro-chemical reactions, the compilation by Poulsen may
be highly recommended ("Botanische Microchemie," Cassel, 1881.
American edition by Trelease, see page 49). Compare also Tschirch,
'Microchemische Reactionsmethoden im Dienste der technischen Micro-
scopie," Archiv der Pharm., 1882.

There is another body very closely related to protoplasm, and like this consisting of protein substances, which is likewise of the greatest importance, pharmacognostically. This is **aleurone**,[1] or protein granules, which are found in numerous seeds of the Umbelliferæ, and Euphorbiaceæ, in *Vitis vinifera, Silybum Marianum, Myristica, Amygdalus*, Cardamomum, and the *Brazil*

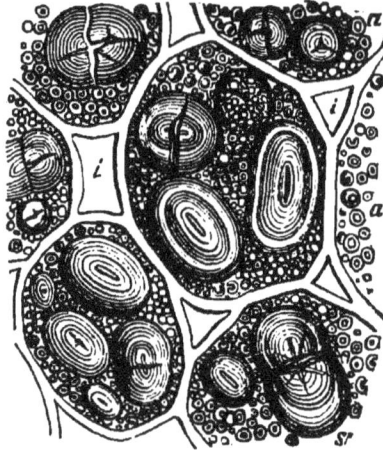

Fig. 29.—Elliptical, plainly stratified starch granules, *s t*, with a broad, central hilum, from the cotyledon of a seed of Pisum sativum, after the addition of water. *a*, protein substances (aleurone): *i*, intercellular spaces (Sachs).

nut (*Bertholletia excelsa*).[2] In many cases the granular contents of the cell, when more strongly magnified, may be resolved into

[1] *Ἄλευρον* the fine flour of grain, gluten (German: *Kleber*) of Hartig, in distinction from amylon. Aleurone was discovered by Hartig (Botanische Zeitung, 1855, p. 881, and 1856, p. 257). For the most thorough examination of it we are indebted to Pfeffer, see Pringsheim's Jahrbücher für wissenschaftliche Botanik, viii. (1872), 429.

[2] The following investigators have contributed to the knowledge of the crystallized, vegetable albuminous bodies : Ritthausen (many publications in the Journ. für prakt. Chemie of the last few years), Maschke, Nägeli, Sachsse, Weyl, Schmiedeberg, Barbieri, Schimper, Drechsel, De Luynes, Grübler (Journ. für prakt. Chemie, 1881); in the latter, the literature is collated. Compare also Husemann and Hilger, "Die Pflanzenstoffe."

7

numerous separate granules of a roundish or polyhedral form [1] (as in the cotyledons of the pea[2] and bean, Fig. 29 *a*), which fill the intervening spaces between the starch granules (in the pea), or the entire cell (the glutinous layer in the seeds of cereals). To these small granules the name of *aleurone* may likewise be applied. In a more restricted sense, the name of aleurone is applied to those large granules which, imbedded in a homogeneous mass of albumen, replace starch granules to a certain extent, and which consist of a fundamental mass of an albumen-like substance, inclosing crystalline (calcium oxalate) or seemingly crystalline, roundish bodies (*globoids*). The albumen-like, fundamental mass is either amorphous or crystalline (*crystalloids*); in the latter case, it is surrounded, together with the inclosed substances, by an enveloping, amorphous mass.

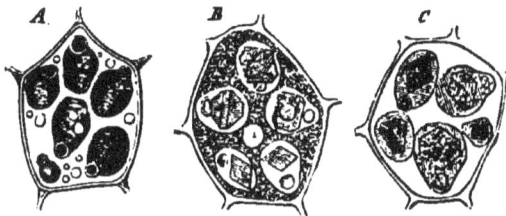

Fig. 30.—Cells from the albumen of *Semen Ricini* (Sachs). *A*, a single cell in concentrated glycerin; the contents show but indefinitely formed masses. *B*, the same section with a little water added, whereby crystalloids, fine granules of protein substances and drops of oil are rendered visible. *C*, the same section warmed with more diluted glycerin, whereby the drops of oil become expelled, and the crystalloids attacked and gradually dissolved.

The globoids (phosphates of calcium and magnesium) are never wanting. Crystalloids[3] occur handsomely developed in

[1] Mounting mediums containing water must, however, be avoided in the preparation, since the granules thereby become destroyed, as has occurred, for instance, in Fig. 29. In the examination of aleurone granules with inclosed substances, concentrated glycerin or fatty oil is always to be employed.

[2] Compare Tangl, "Das Protoplasma der Erbse." Sitzungsbericht der Wiener Akademie, 1887.

[3] From κρύσταλλος crystal, and εἶδος similarity. They owe their name to C. Nägeli, "Sitzungsber. der Münchener Akad.," 1862, p. 121.

the aleurone granules of the seeds of *Elœis guineensis, Æthusa Cynapium*, and all Euphorbiaceæ (*Ricinus, Croton*); they are wanting in the aleurone granules of umbelliferous seeds. Crystalloids occur, together with crystals, in *Æthusa Cynapium*.

Occasionally an aleurone granule in each cell is distinguished from the others, either by its size alone, or also by inclosing crystals of a different formation or larger size (Fig. 31 *A* at *c*). Such a granule is termed solitary (German, *solitär*, Hartig).

The *crystalloids* are doubly refractive,[1] their angles, however, are inconstant; they are insoluble in water. The aleurone granules free from crystalloids, on the contrary, dissolve for

B   A   C

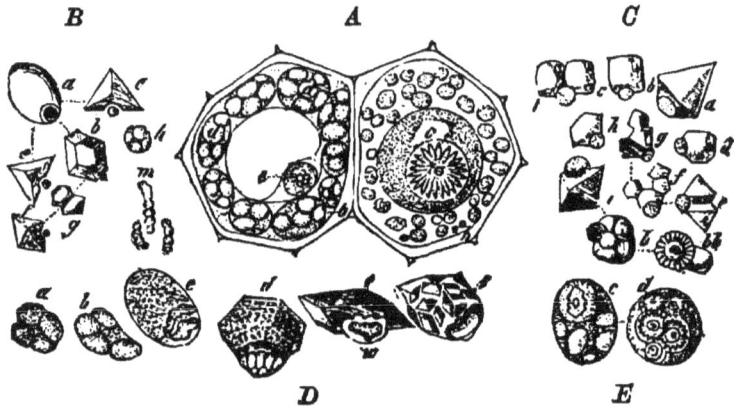

D   E

FIG. 31.—*A*, Two gluten-cells from the seed of the *raisin*. In the cell at the left much granular protoplasm and a nucleus *c* is present. The cell at the right, after complete ripening, with a large, solitary granule (*c*), and numerous, small aleurone granules. *B*, Aleurone from the seed of *Ricinus communis* with crystalloids. *C*, Aleurone from *Euphorbiæ, Myristica* (*c*), *Croton* (*b*), *Phyllantus* (*b b*). *D*, Aleurone from the seed of *Bertholletia excelsa*, *f* dissolution of a crystalloid into several crystals. *E*, Aleurone from the seed of *Lupinus* (*c*) and *Conium* (*d*) (Hartig).

the most part in pure water (*Pæonia, Lupinus*), and all of them in feebly alkaline water. The fundamental mass, consisting of protein substances, is insoluble in alcohol, ether, benzol, chloroform, and paraffin; it is colored yellow by iodine.

[1] They therefore appear more clearly in polarized light. Compare also Radlkofer, "Krystalle proteinartiger Körper," Leipzig, 1859.

The *globoids* dissolve in inorganic acids, also in acetic and tartaric acids, but not in a dilute solution of potassa.

There occur, moreover, in the vegetable kingdom, crystalloids which are not inclosed in aleurone granules (the *potato*, Fig. 108 *a*).

The aleurone granules, as is already indicated by their exclusive occurrence in seeds, belong to the reserve substances which have the function of presenting to the germinating plant, in its first development, and before it is capable of assimilating independently, sufficient material for building up its organs. They are, therefore, of the greatest importance in the economy of the plant. How abundantly protein substances are contained in some seeds is shown by the following figures, showing their percentage : *Nux vomica*, 11 ; *Cacao*, 13 ; *Black Mustard*, 18 ; *Almond*, 24 ; *Linseed*, 25 ; *Ignatia seed* and *White Mustard*, 27.

These numbers are calculated from nitrogen estimations, with the presupposition that albuminous bodies contain 15 per cent of nitrogen.

With the albuminous bodies are directly connected the **chlorophyll bodies,** the fundamental mass of which likewise consists of an albumen-like substance.[1] This fundamental mass (stroma), which, moreover, is also very soft, is of a sponge-like structure (Fig. 33 *a*), and contains in the meshes of the frame-work a small amount of the mixture of coloring substances, to which Pelletier and Caventou[2] in the year 1817 gave the name "chlorophyll."[3] The crude chlorophyll consists of two coloring matters, chlorophyll in a more restricted sense, or pure chlorophyll,[4] and xanthophyll.[5] The former is bluish-green, the latter yel-

[1] Sachs, "Flora," 1862 and 1863.—Mohl, "Verm. Schriften."

[2] Journ. de Pharm., 1817, p. 486.

[3] Χλωρός green, and φύλλον leaf.

[4] Compare Tschirch, "Untersuchungen über das Chlorophyll," Berlin, 1884. In this publication, the entire literature relating to chlorophyll to the year 1883 is critically sifted and reviewed. At the close of the work, a catalogue of the literature is given, comprising nearly 600 investigations.

[5] Ξανθός yellow, and φύλλον leaf.

ow.   The emerald-green color of leaves is thus a mixed color,[1] and the spectrum of the leaf a mixed spectrum.   While chlorophyll only presents bands in the less refractive half of the spectrum (red-green), and shows a continual absorption of the violet : in the case of xanthophyll there occur no bands at all (Fig. 32 *s*) between red and green, but only in the blue.   Pure chlorophyll can be prepared, as one of us (T.) has shown, by the reduction of chlorophyllan, a crystallizable body.

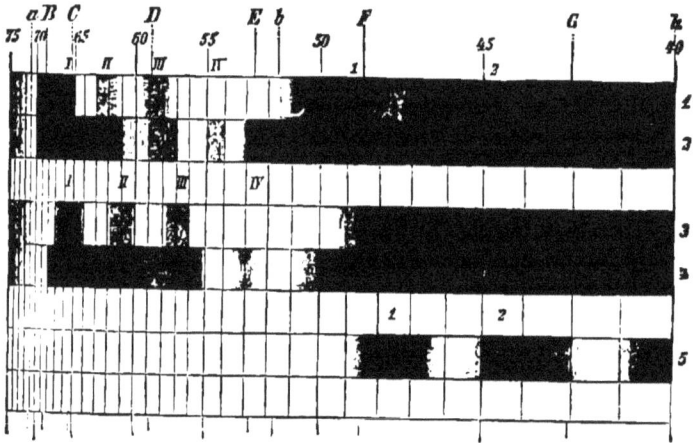

FIG. 32.—1. Spectrum of 2 leaves.
2. Spectrum of 5 leaves.
3. Spectrum of a dilute | alcoholic solution of pure chlorophyll.
4. Spectrum of a concentrated |
5. Spectrum of an alcoholic solution of xanthophyll.
  In the leaf spectrum, band 2 of the xanthophyll is mostly covered by the projecting terminal absorption of the pure chlorophyll; at least it is always rendered unclear (Tschirch).

The chlorophyll granules of the higher plants always appear as roundish, disk-like bodies (Figs. 33, 109, 129, 161), which,

[1] By means of benzol, as has been shown by G. Kraus ("Zur Kenntniss der Chlorophyll-Farbstoffe," Stuttgart, 1872), an alcoholic tincture of chlorophyll from leaves may be split into two layers, a yellow lower layer containing xanthophyll, and a green upper layer which contains the chlorophyll.   The separation is, however, not quantitatively exact. (Tschirch, "Untersuchungen über das Chlorophyll," Berlin, Parey, 1884.)

when they lie close beside each other, become flattened polyhedrically, without, however, coming in contact, for the reason
that they are provided with a thin membrane of protoplasm.
They are the organs in which the most important process of
plant life is effectuated, *viz.*, the assimilation of carbonic acid
under the influence of light, with the formation of organic or
carbon compounds. Only organs containing chlorophyll are
capable of effecting this change. Indeed, we also find in the
chlorophyll granules an abundant accumulation of assimilation
products, and especially starch (Fig. 33, *b, c, d*). If a leaf of
the peppermint, after the coloring matter has been extracted by

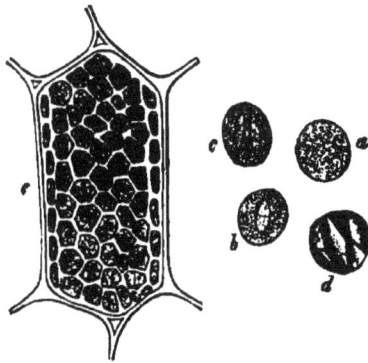

Fig. 33.—*a*, Chlorophyll granule, the sponge-like structure indicated by punctations; *b, c, d*, inclosures of starch in the chlorophyll granule; *e*, a cell with chlorophyll granules located along the wall (Tschirch).

alcohol, is placed in iodine-water (see Micro-chemical Reagents),
it assumes at one a bluish-black color; every chlorophyll granule contains some starch granules of a black color (see the
iodine-starch reaction).

The chlorophyll granules always lie imbedded in the protoplasm-sac, on the inner wall of the cell, and shrink by the
contraction of the sac, by the addition of reagents, or by the
death of the cell. Since the fundamental mass of the chlorophyll
granules is very soft, many of these flow together by this process
to form larger masses. Thus in drugs (green leaves and stems)

the chlorophyll granules are seldom found unchanged, mostly forming within the cells shapeless masses, in which the granular structure can be recognized only with difficulty.

A similar condition also exists with regard to the maintenance of the coloring matter, the chlorophyll.[1] If, namely, leaves are quickly dried, the plant acids act but slightly upon the chlorophyll, only a little brownish-yellow chlorophyllan (an oxidation product of chlorophyll) is formed, and the leaves remain handsomely green. If, however, they are dried without care and slowly, brownish-yellow leaves are obtained, in consequence of the abundant formation of chlorophyllan.[2] Some leaves, however, become brown even with the most careful drying (*Nicotiana, Juglans*).

Since the formation of chlorophyll is dependent upon light, it is found only in those parts of plants which grow above ground and are exposed to the light.[3] Leaves developed in the dark are yellow (*"etiolement"* etiolation ; the coloring matter is called etiolin[4]). All leaves and green shrubs contain chlorophyll, although it is sometimes concealed by red coloring matters dissolved in the cell-sap (*Dracæna* leaves). We meet with it also in seed-vessels (*Juglans*), and in barks, especially in the thinner ones (*Rhamnus, Salix, etc.*). Since, however, it occurs only in cells which still possess the functions of life, it is wanting in such barks as consist entirely of permanent tissue, or in which the peripheral layer is wanting (*Cinchona, Cinnamon*).

The chlorophyll coloring matter, being a harmless green color, is of practical importance.

[1] The name chlorophyll must remain confined to the coloring substance.

[2] These circumstances have been thoroughly considered by Tschirch: "Einige practische Ergebnisse meiner Untersuchungen über das Chlorophyll," in Arch. der Pharm., 1884.

[3] Nevertheless, the half-underground leaf-bases of *Rhizoma filicis* are also green. Exceptions are presented also by many green embryos enclosed by untransparent fruit-casings, and by the small embryos of the Coniferæ, developed in the dark.

[4] From the French word *étioler*, to etiolate or become blanched, which is derived from the latin *stipula*, halm.

The chlorophyll of leaves (crude chlorophyll) is insoluble in water, but soluble in alcohol, ether, carbon bisulphide, acetone, benzol, volatile and fatty oils (*Oleum Hyoscyami* of the Pharmacopœa Germanica is colored by chlorophyll), chloroform, and dilute solutions of caustic potassa (in the latter, with chemical change), forming emerald-green solutions, which are dichroic (green-red), and also show a magnificent fluorescence.

A convenient method for distinguishing chlorophyll from other green coloring matters, is as follows : The alcoholic solution of the coloring matter is shaken with concentrated hydrochloric acid and ether, the acid solution then becomes blue, the ethereal yellow. No other green coloring matter shows precisely the same deportment. Pure chlorophyll dissolves with a blue color in hydrochloric acid, and is soluble in the same solvents as crude chlorophyll (see above). Pure chlorophyll appears to stand chemically in close relation to the lecithines, or to be itself a lecithine.

The **colored crystalloids** of many flowers and fruits (*Capsicum annuum*, *Rosa*, *Crocus*, *Carthamus*, *Tropæolum*, *Chrysanthemum*) should also be considered here. The development of these proceeds mostly in such a manner that the chlorophyll bodies—in the beginning the flowers and carpels of the fruit are mostly green—by a disturbance of their form and loss of their original color, pass into the crystalloid coloring matters. The yellow coloring matters (anthoxanthin [1]) especially occur often in the form of handsomely developed crystals (Fig. 34), as in the Carrot, and they probably always possess, besides the coloring matter, a plasmatic basis.[2] Occasionally these coloring matters also appear in the form of granules.

The **red and violet coloring matters** (anthocyan [3]) are, as a

[1] From ἄνϑος flower, and ξανϑός yellow.
[2] They are capable of swelling. These crystalloids have recently been repeatedly examined, thus by Hildebrand, Pringsheim's "Jahrbücher," 1861.—Nägeli, "Sitzungsberichte d. Münch. Akad.," 1862.—Weiss, "Sitzungsber. d. Wiener Akad.," 1866.—Schimper, Botan. Zeit., 1883.—A. Meyer. *Ibid.*, 1883.
[3] From ἄνϑος flower, and κυάνεος blue.

rule, dissolved in the cell-sap (red potatoes, red foliage leaves and petals).

In the fundamental protoplasmic mass, there is very frequently found liquid or solid **fat,** for instance, in the embryo of the Gramineæ, in the endosperm of *Ricinus,* and in the cotyledons of the Cruciferæ. The fat appears to be combined with the protoplasm in the finest state of division. Microscopical sections of seeds which are very rich in fatty oil (*Ricinus, Tiglium, Amygdalus, Corylus*) show, when observed under water, a number of small oil-drops, which are not visible when alcohol or glycerin, instead of water, is used as the mounting medium. It is only after gradually diluting the alcohol or glycerin under the cover-

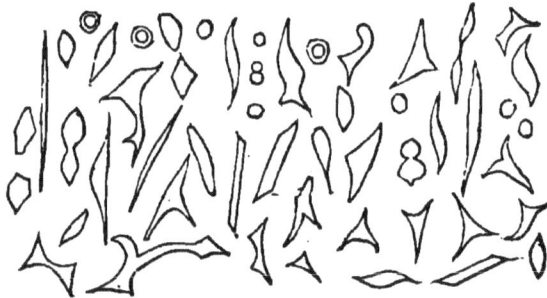

Fig. 34.—Crystalloid coloring matters (anthoxanthin bodies) from flowers and fruits Tschirch).

glass with water, that the oil-drops are brought to view. From this it may be concluded that the fatty oil is contained in the dry seed in combination with another substance, which prevents the oil from flowing together in drops. This evidently very loose compound (perhaps containing albumen) is destroyed by water, and the oil then unites in the form of drops.

This result may, however, be so interpreted that the fatty oil occurs in the cells very intimately mixed with the protoplasm, that the albuminous body mixed therewith becomes dissolved by the water, and that thus the oil is caused to form larger drops. However this may be, the fatty oil is evidently very effectually protected by this highly remarkable manner of its storage ; it is, indeed, sufficiently well known that the oil becomes quickly rancid when the seeds are comminuted or even moistened.

Many cells of other tissues contain, moreover, free fat in a liquid or solid form. In the former case, the drops of oil admit of especially easy recognition on account of their remarkable refraction of light, e. g., in *Secale cornutum*, and in *Senega root*. The fats deposited in a solid form are crystalline, which may be seen with special clearness, among other examples, in *Cacao, Cocculus Indicus*, and in the *Nutmeg*.[1] The fat contained in the kernels and shells of the *Cocculus* fruits consists almost entirely of free stearic acid.[2]

In *Stillingia sebifera* (Nat. Ord. Euphorbiaceæ), there is found upon the surface of the black seeds a coating of fat, and in *Peckia (Cybianthus) butyrosa* (Nat. Ord. Myrsineaceæ), each of the four nuts has a pericarp several millimetres in thickness, the inner portion of which forms a yellow, leafy substance.

Fats are found, not only in the seeds, but occasionally also in the fleshy portion (sarcocarp) of fruits (a great deal in that of the Oil-palm, *Elæis guineensis*, the Japanese wax-trees *Rhus succedanea* and *Rhus vernicifera*, the olive *Olea europœa*), in pollen, spore-cells (*Lycopodium, Pollen Pini*), in some roots (*Cyperus esculentus*), and in the passive state of fungi (*Secale cornutum*).

As has already been mentioned, fats and oils are found in small amounts in almost all tissues which exercise the functions of life ; they occur regularly in seeds. This is readily seen when a section is treated with concentrated sulphuric acid ; the protoplasm and membrane become immediately destroyed, and the small drops of oil, which are otherwise scarcely visible, flow together to form larger drops, which are not attacked by the sulphuric acid.

This is, in general, the best method for the detection of small amounts of fatty oil in microscopical preparations. In this

[1] Compare also Möller, "Ueber Muscatnûsse." Pharm. Centralhalle, 1880, No. 51-53.

[2] Schmidt and Römer, Archiv der Pharm., 221 (1883), 34.

way, one may readily succeed in rendering visible, for instance, the fatty oil of *Lycopodium*, of which the latter contains nearly 50 per cent, but which does not admit of recognition by simple microscopical observation ; it is only necessary to crush the grains, and then to add the sulphuric acid or concentrated solution of calcium chloride.

In seeds, the fats play the part of reserve substances, and in living tissues, particularly those containing chlorophyll, they are an assimilation product, which evidently finds at once further application in building up the tissues.

The fats are soluble in boiling alcohol, in ether, carbon bisulphide, benzol, paraffin and volatile oils, and are colored brownish-black by osmic acid.

With those seeds which contain fatty oil most abundantly, this may exceed half the weight of the kernels (after the removal of the seed-shells). Thus in *Amygdalus, Cacao, Papaver, Ricinus, Sesamum,* and *Croton Tiglium ;* in the latter, the oil amounts to nearly 60 per cent. For the most part, however, the amount of fat of other seeds which concern us here is small ; *linseed* and *black mustard* afford about 33 per cent of oil.

The fat of the *olive*, the yellow *palm-oil*, as also the so-called *Japan-wax*, are contained in the fleshy portion (sarcocarp) of the fruits of the respective plants from which they are derived. The remaining solid and liquid fats of the vegetable kingdom, which are brought in large amounts into the markets of the world, are furnished by seeds.

The fats are esters (compound ethers) of propenyl or glycerin. The acids combined with this radical belong mostly to the series of the ordinary fatty acids, although a portion of very many fatty oils and even of solid fats consists of olein, *i. e.*, of the propenyl ester of oleic or elaïc acid, which belongs to the acrylic acid series. Nowhere has a propenyl ester been proved to exist singly in nature ; every fat is a mixture of several such esters. When a fat is decomposed (saponified) by means of a caustic alkali, the base is therefore always found to be combined with more than one acid.

For us the most remarkable and important constituent of the cell-contents is the **starch** (amylum [1]).

The latter occurs abundantly, in the form of characteristic granules, in seeds and other receptacles of reserve substances (rhizomes, tubers). The seeds which are provided with starch (reserve-starch) are, however, very much less numerous than those which contain none. It appears also in the conducting tissues (transitory starch), and in the interior of the chlorophyll granules (assimilation-starch, autochthonous starch), but then mostly in very small granules. For our purpose, the starch granules of the reserve-receptacles are especially important.

Between the starch and other constituents of the cells there exist manifold, but as yet only slightly explained relations. Thus in the case of *Radix Belladonnæ*, Budde[2] has found certain relations to exist between the amount of starch contained in the root and the amount of alkaloid. The amount of atropine is most considerable in roots which are very rich in starch, and least in those which are free from starch (compare also page 13).

Starch is organized and appears in the form of more or less distinctly stratified granules[3] (Figs. 35, 36, 37, 39, 42, 43, 44, 45, 46).

Some drugs are exposed, in their fresh condition, to a higher temperature in order to dry them more quickly. If these parts of plants are juicy, the amylum thereby suffers that change which is known as the formation of paste. The granules swell to a high degree and flow together, to form structureless masses or balls of paste. Thus in the case of *Curcuma*,[4] *Jalap*, *Salep*, some varieties of *Sarsaparilla*, and the East Indian *Aconite* tubers. *Sago* is nothing more than swollen and dried starch.

The *layers* (which are especially handsome in the granules from the potato and leguminous seeds) are arranged around a

[1] From α (α privative), and μύλη mill—flour prepared without a mill.
[2] Archiv der Pharm., 220 (1882), 414.
[3] Nägeli, " Die Stärkekörner, Pflanzenphysiologische Untersuchungen," 1858. The most comprehensive work relating to starch.
[4] Berg, "Anatomischer Atlas," Taf. xix., Fig. 48.

common central point of the granules, which, in consequence of unequal growth, are mostly not uniformly round ; in granules of very eccentric construction, however, the layers are in the form of immeasurably thin shells on the side having the slightest growth. The layers (Fig. 35) originate through an abrupt variation in the amount of water of the separate zones. An outermost layer containing but very little water is followed by one with an abundance of water, then again by one poor in water, *et cetera.* The centre of the granule, the *nucleus,* is very rich in water.

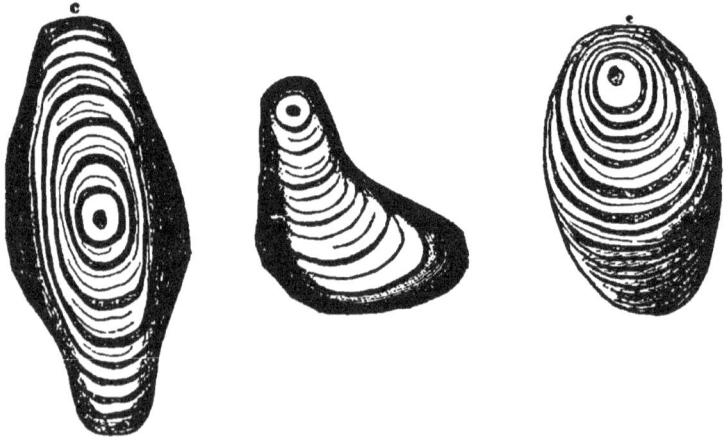

FIG. 35.—Starch granules with very distinct layers and hilum, from the potato, very highly magnified.

If the starch granule advances no farther in growth, a cavity (hilum) generally remains in place of the nucleus. This space is often confined within a very small compass, and therefore appears as a small, dark point (nucleus-point) in the starch of the potato and of the rhizomes of some Ziugiberaceæ (Figs. 36 and 46). In the starch granules of *Tuber Colchici, Maranta* (Fig. 45), *Maize* (Figs. 48 and 49), *Radix Calumbæ* and others, the somewhat larger hilum often assumes the form of a star or a cross (Fig. 37), and in many seeds from the family of Leguminosæ, as in *Semen Calabar* and the *Bean,* the hilum is proportion-

ately very wide, and extended in the direction of the axis of the
frequently elliptical granules (Fig. 44).

The nucleus, as a rule, is located eccentrically, although it is
central in the large granules of the cereals (Fig. 47) and in the
small, round granules of very many plants.' Occasionally several nuclei are found in one granule.

The layers disappear in consequence of the abstraction of water,
when the granules are observed under liquids free from water,
such as benzol, paraffin, volatile oils, fatty oils, glycerin, or when
they are warmed.   Glycerin loses this property to a degree proportionate to the amount of water it contains.   On the other
hand, the distinction of the layers is also obliterated through
their intumescence (by the addition of water), even by water
at 60 to 70° C. or a still higher temperature, but even in the

FIG. 36.   FIG. 37.

FIG. 36.—Starch granules from the rhizome of ginger (Hager).
FIG. 37.—Starch granules with a star-shaped hilum, from *Tuber Colchici.*

cold by means of saturated solutions of many bodies which are
very readily soluble in water, such as caustic potassa or soda,
potassium iodide, calcium chloride, sodium nitrate or acetate,
and chloral hydrate.   These substances increase the capacity of
absorption of water by the starch to an enormous degree, far
beyond the distinction of the separate layers just explained, so
that these swell up to a uniform mucilage.

If starch granules are pressed under the cover-glass, fissures
and clefts are formed, which, proceeding from the cleft of the
nucleus or from the periphery, form cracks having a course
mostly at right angles with the layers.

According to the very thoroughly founded and developed

' The eccentricity in the case of *Cyperus esculentus* amounts to ⅓,
in *Canna lanuginosa* to 7/16.

views of Carl Nägeli,[1] starch grows in such a manner that the
formative material inserts itself between the layers of the
granule, and is by no means added externally through "apposi-
tion." The objections raised in opposition to Nägeli's view of
"intussusception,"[2] especially by Schimper[3] and by Arthur
Meyer,[4] are deduced from the supposition that starch possesses a
crystal-like character. Its crystalloids, as in other carbo-hydrates
(see text to Fig. 54 *a* and *b*), are united in the form of spheres,
sphæro-crystals, but are highly characterized by the capability of
swelling. Their stratification is the result of alternate solution
and renewed deposition of solid substance. That the granules
are less dense toward the interior is shown by the penetration
of the solvent.[5]

Occasionally two nuclei surrounded by a separate series of

FIG. 38.   FIG. 89.

FIG. 38.—Amylum, compound granules with a common integument (Dippel).
FIG. 39.—Compound starch granules from *Radix Sarsaparillæ*.

layers (Fig. 38) are formed in a single starch granule ; if these
nuclei continue to separate from each other, a high tension is
produced in the layers common to both, which leads to the dissolution of the double granule into two separate ones (fractured
granules). If, instead of two, a still larger number of granules
appears, *compound granules* are produced, which may consist of

[1] "Die Stärkekörner," Zurich, 1858. Large octavo, 624 pages and 10
plates (mentioned also on page 108, foot-note 3).—" Sitzungsberichte d.
Münch. Akad.," 1863 and 1881; Bot. Zeitung, 1881, 633; also Nägeli and
Schwendener, " Das Microscop," 1877, p. 423.

[2] *Intus*, in or within ; *suscipere*, to take up.

[3] Botanische Zeitung, 1881, 185.

[4] *Ibid.*, 1881, p. 841, and 1884, p. 508.

[5] Compare further the respective references thereto in Just's Bot
Jahresbericht for 1881, I., 398-400.

very numerous individual granules (*Avena*, Fig. 40, *Spinacia, Sarsaparilla*, Fig. 39).

The *false compound granules* are formed by several separate granules becoming firmly agglutinated with each other through mutual pressure (frequently occurring in the starch in chlorophyll, Fig. 33 *d*).

The shape of starch granules is very varied.[1] The fundamental form is the sphere. All of the smaller, isolated starch granules possess this form, such as the small granules of the *wheat* (Fig. 47), of the *potato* (Fig. 43), and the so-called transitory starch. When the granules fill the cell and are densely crowded, they are always flattened through mutual pressure, and are then mostly polyhedric (dodecahedron and

 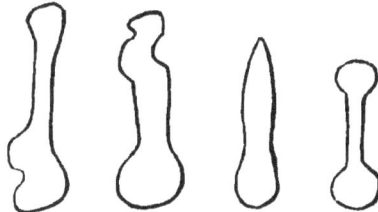

FIG. 40.   FIG. 41.

FIG. 40.—A compound starch granule of the *Oat*, resolved into its separate granules.
FIG. 41.—Starch in bone-shaped and club-shaped granules from the milky juice of *Euphorbia antiquorum*; more difficult to obtain from the commercial '' Euphorbium '' (gum-resin of *Euphorbia resinifera*).

allied forms) (*Maize*, Fig. 49, *Rice*, Fig. 50). In the chlorophyll granule the amylum is mostly spindle-shaped (Fig. 33 *c*, *d*). Club-, staff-, or bone-shaped structures are found in the milky juice of many *Euphorbiæ* (Fig. 41), club-shaped in the rhizome of *Galanga*,[2] and branched in the root-stock of *Nelum-*

[1] Compare the illustrations and descriptions of forms of starch by Vogl, '' Die gegenwärtig am häufigsten vorkommenden Verunreinigungen, etc., des Mehles,'' Vienna, 1880.—R. von Wagner, '' Die Stärkefabrikation.'' Braunschweig, 1876.—König, '' Die menschlichen Nahrungs- und Genussmittel,'' Berlin, 1883.—F. von Höhnel, '' Die Stärke und die Mahlproducte, etc.,'' Cassel and Berlin, 1882.

[2] Berg, '' Anatomischer Atlas,'' Plate xix., 46.

bium speciosum, Willd. *Sago-starch*[1] is provided with swollen protuberances (Fig. 42). Independent of exceptions of this character, the spherical and ovate, or often flattened forms[2] predominate.

Although not susceptible of strictly mathematical definition, the size and shape of the starch granules are nevertheless characteristic for individual plants. A knowledge of these peculiarities is therefore indispensable in the examination of flour and varieties of starch. Besides the smaller structures, which usually occur, each granule possesses a predominating typical form. It is only when the latter has been confirmed, in its definite shape and size,[3] by a large number of granules, that one can assume that a certain variety of starch is present.

In the examination of starches, this is the only means for

FIG. 42. FIG. 43.

FIG. 42.—Starch granules of Sago (Hager).
FIG. 43.—Potato starch (Koenig). *a*, the nucleus. Compare also Fig. 108.

their identification, but it is otherwise with varieties of flour. The latter consist of ground fruits and seeds, and thus contain, besides the starch granules, remnants of cells of the inner tissue as well as of the integuments, and often also remnants of hairs (*Triticum*). In the case of flour, these remnants may therefore be very well used as a guide in their examination.[4]

[1] Compare also Wiesner, " Die Rohstoffe des Pflanzenreiches," Leipzig, 1873.

[2] For example in Zingiber, Berg's Atlas, Plate xx., 49.

[3] Measurements of the size of the granules (by the aid of the ocular micrometer) must always be undertaken. The linear diameter is determined. The unit of measure is the micro-millimeter (μ or mic.) = $\frac{1}{1000}$ mm. = 0.000001 m.

[4] Compare Wittmack, " Anleitung zur Erkennung organischer und

8

The most important forms of starch are the following :[1]

1. POTATO STARCH (*Solanum tuberosum*, Figs. 43 and 108). Type : large, eccentric, very distinctly stratified, quite irregular granules, which are rounded by three or four corners ; they are often rhombic and wedge-shaped, but never flattened. The nucleus is at the smaller end.

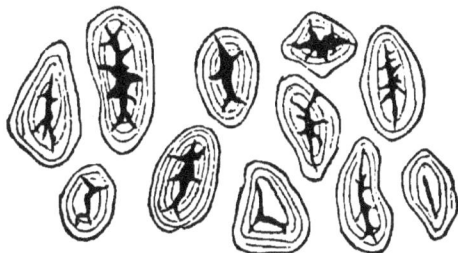

FIG. 44.—Bean starch (Tschirch).

Secondary form : small, roundish, and medium sized, half or completely compound granules.

FIG. 45.—Maranta starch (Tschirch).

2. BEAN STARCH (*Physostigma, Vicia* and species of *Phaseolus*, Fig. 44). Type : bean-shaped to kidney-shaped, distinctly stratified granules, which are always simple. The

anorganischer Beimengungen im Roggen- und Weizenmehl," Leipzig, 1884.

[1] A good key for use in determining the varieties of starch has been given by Vogl : " Nahrungs- und Genussmittel aus dem Pflanzenreich," Vienna, 1872 ; compare also König, " Nahrungsmittel," II., p. 405, Wagner and others.

nucleus is not visible, since the granules are nearly always traversed by a broad, radiately branched, longitudinal cleft.

Secondary form : small, roundish granules.[1]

3. MARANTA STARCH, Arrowroot starch (*Maranta arundinacea*, Fig. 45). Almost entirely typical forms : more or less flattened, nearly quadrangular, rhombohedral, triangular, club- or pear-shaped granules, with a distinct nucleus located at the broad end, and often traversed by an occasionally three-rayed, lateral cleft. Stratification distinct.

4. EAST INDIAN ARROWROOT (*Curcuma leucorrhiza* and *C. angustifolia*, Fig. 46). Type: distinctly stratified, flat, tabular or oval and lengthened, triangular granules, drawn out to a point on one side, which contains the nucleus, without a cleft.

Secondary form : small, triangular granules.

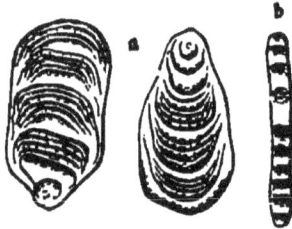

FIG. 46.—Starch of *Curcuma leucorrhiza* (Koenig).

5. WHEAT STARCH (*Triticum vulgare*),[2] Fig. 47. Type:

(*a*) LARGE GRANULES, flatly lenticular, almost exactly circular, without a cleft or a distinct nucleus. They are four times larger than the following :

(*b*) SMALL GRANULES, roundish or polyhedral, often connected in pairs.

[1] Specific characters for the discrimination of bean- and pea-starch have been given by Tschirch, " Stärkemehlanalysen," in Archiv der Pharm., 222 (1884), p. 921.

[2] The rye and barley have similar granules. For further details regarding them, compare the previously mentioned monographs on starch.

Secondary form : slightly roundish granules, having a form intermediate between the two preceding.

6. MAIZE STARCH (*Zea Mays*, Figs. 48 and 49) presents only typical forms.

(*a*) Horn endosperm (Fig. 49): sharply angular granules,

FIG. 47.—Wheat starch; *v*, face view ; *s*, marginal view ; *t*, parts of a double granule : *r*, a granule of rye starch with a three-rayed cleft (Tschirch).

without distinct stratification, and mostly provided with a

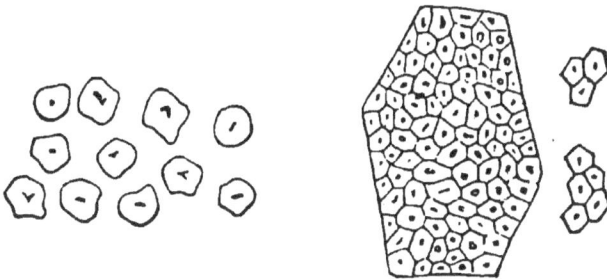

FIG. 48.   FIG. 49.

FIG. 48.—Maize starch granules from the farinaceous endosperm (Tschirch).
FIG. 49.—Maize starch from the horn endosperm (Tschirch).

cleft ; they are in close contact with each other, and completely fill the cell. In the flour also, several granules are often still coherent.

(*b*) Farinaceous endosperm (Fig. 48): granules more round-ish, occasionally without a cleft. They do not completely fill the cell, and therefore are not sharply flattened polyhedrically against each other.

7. RICE STARCH (*Oryza sativa*, Fig. 50), only typical forms ; very sharply angular, almost crystal-like, fractured granules, occasionally several still connected, without a nucleus cleft.

FIG. 50.—Rice starch (Tschirch).

8. OAT STARCH (*Avena sativa*, Fig. 51). Type : large, oval aggregations, as much as $\frac{50}{1000}$ mm. (50 μ) in size, composed of from two to three hundred granules and their components (Fig. 51 *b*). The latter are polyhedric and sharply angular, without a distinct nucleus.

FIG. 51.—Oat starch. *a*, Secondary form—filling granules ; *b*, component granules of the aggregation (Tschirch).

Secondary form : small, roundish, spindle-shaped,[1] similar to the fractured granules, the so-called " filling starch."

Like the cell-membrane, the starch granules, in consequence

[1] Moeller (" Die Mikroskopie der Cerealien," in the Pharm. Central-halle, 1884, No. 44–48) declares these to be the characteristic forms.

of their stratified structure, are also doubly refractive. In polarized light each granule displays a black cross, the arms of which intersect at the hilum (Fig. 52).

When the structure of the granule is destroyed, either through tumefaction or by torrefaction, it immediately loses its optical properties, though agents which produce swelling, but which have neither an acid nor an alkaline reaction, produce no change, at least in a chemical sense, in the substance of the starch. The optical properties therefore appear to be dependent upon the manner of construction of the granule. Nevertheless, Nägeli entertains the view that the *micellæ*[1] of starch (a term which he applies to the complex of atoms, surrounded by a film of water, which through intussusception become separated from each other), like those of the denser cell-membranes, are originally crystalline, and show the deportment of optically uniaxial crystals, but only so long as they have not become disintegrated, which Nägeli assumes to take place upon swelling.

Starch, as the most important reserve nutritive substance, is contained in an extraordinarily large number of reserve receptacles, thus in seeds (endosperm of the cereals, the cotyledons of many Leguminosæ), *Semen Cacao, Sem. Myristicæ, Sem. Paradisi, Sem. Piperis* (v. *Piper album*), *Sem. Quercus,* while *Sem. Cydoniæ, Sem. Lini, Sem. Sinapis albæ* and probably others contain starch, at least before ripening. Furthermore, in rhizomes (*Maranta,* the Zingiberaceæ, *Aspidium Filix-mas, Asarum, Calamus*), in roots (*Althæa, Sarsaparilla, Krameria, Ipecacuanha, Rhubarb, Belladonna*), and in tubers (*Potato, Salep, Jalap, Colchicum*).

A remarkable exception is presented by the roots and rhizomes of the Compositæ, which contain no starch, or only transitorily, and then but extremely small amounts of it. It is furthermore wanting in *Radix Gentianæ, Rubiæ, Saponariæ, Senegæ,* and in the rhizome of *Triticum repens,* at least in the stages of development which here come under consideration. In all these organs its place is supplied by other substances.

---

[1] Unjustifiable diminutive of *mica*, a small crumb.

Ultimately, the entire starch of a plant owes its origin to the chlorophyll granules, and although it must be accepted with certitude that it is not the first product of assimilation, it is, nevertheless, the first which is visible. From the assimilating tissue it then emigrates into the conducting tissues, and in these to the place of consumption (constructive tissue), as also into the repositories of the reserve substances (seeds, rhizomes).

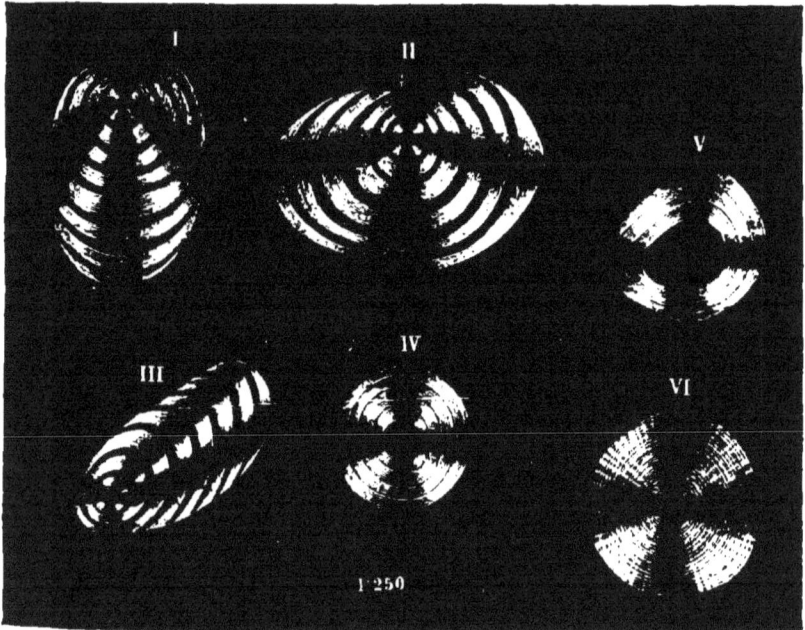

FIG. 52.—I. to V. Starch granules in polarized light; all the layers are traversed by a dark cross proceeding from the organic centre. VI. Inulin (Dippel).

Since it cannot, however, as such, penetrate the membranes, it is very probably converted previously, through fermentative action, into dextrin or sugar. These substances can circulate diosmotically through the membranes. Nevertheless, the "paths of the starch"[1] are also characterized by the occurrence of small

[1] In German, "Stärkebahnen."

starch granules, since the above-named transformation products possess the inclination, under suitable conditions, to again become deposited as solid starch (transitory starch). These small starch-granules, which are in process of migration, are thus found in the so-called *starch-sheath*, and in the sieve-tubes and medullary rays. They are also contained in some fruits before ripening (*Olives, Fructus Conii, Fructus Juniperi,* likewise the *Fig*).

How the formation of starch is effected in the chlorophyll granules (assimilating-starch, p. 108) is, meanwhile, still an enigma. Only so much is certain, that for its production light is required,[1] while potassium also appears to be indispensable for it.[2]

The manner of circulation of starch may be elucidated by a single example. The species of *Orchis* which afford *salep* possess, after the close of the period of vegetation, a tuber which is filled with starch and mucilage. The tuber is quiescent during the winter, and in spring develops a stem bearing the leaves and flowers. During the entire first period of development, the tuber provides the young shoot with nutritive material ; the starch migrates from the tuber upward into the shaft. In the course of further development, the leaves unfold, and now undertake on their own part the new formation of starch. But even now the plant provides for a future year. Beside the old, and now entirely exhausted tuber, a new rudimentary one is formed, into which the starch formed in the leaves migrates downward, in order to furnish the constructive material for the young plant during the next year.

If the starch is dissolved in a reserve-receptacle, the granules do not disappear at once, but solution gradually takes place, whereby a peculiar corrosion often occurs. Such *corroded*

[1] Compare Sachs, " Experimentalphysiologie der Pflanzen," 1865 ; furthermore, Botan. Zeit., 1862 and 1864, " Flora," 1863, and researches of the Botanical Institute of Würzburg. 1884.—Godlewski, Krakauer Akadem., 1875.—Böhm, Botan. Zeit., 1876, and others.

[2] Proved by Nobbe in 1871.

starch-granules are particularly well observable in the first stages of germination (Fig. 53).

We are acquainted with starch in the plant only in the solid form, although compelled to assume that it is formed, or in a manner crystallizes out, from a liquid.

Starch forms a glistening powder, the specific gravity of which varies according to its origin, but does not deviate much from 1.5. In the air-dry condition, it incloses from 13 to 17 per cent of water, after the removal of which its density increases, according to its derivation, to from 1.56 to 1.63. While air-dry starch floats upon chloroform, it sinks therein after having been deprived of its water by heating to 100° C. Dried starch quickly absorbs again from the air the eliminated water.

The small amount of incombustible substances which it con-

FIG. 53.—Corroded starch-granules from the endosperm of a young maize plant 8 centimeters (about 3 inches) high, in process of solution. *a*, a granule still intact. (Tschirch).

tains, about 0.5 per cent, can probably be explained only by supposing them to be deposited mechanically.

The composition of anhydrous starch corresponds to the formula $C_{12}H_{20}O_{10}$, although Musculus (1861 and 1870) and W. Nägeli have shown that the formula $C_{18}H_{30}O_{15}$ or $C_{36}H_{62}O_{31}$ corresponds still better to the facts.[1]

Leuchs (in 1831) found that starch granules are attacked by saliva. C. Nägeli, who pursued the subject further, came to the conclusion that the granule is built up from *cellulose* and a peculiar starch substance, *granulose*. According to his view, the

[1] For information relating to the elementary composition of starch, we are indebted to W. Nägeli, Sachsse, Pfeiffer, Tollens and Salomon. See also Husemann–Hilger, " Die Pflanzenstoffe."

saliva acts upon the latter substance, dissolving it, and leaving
a frame-work or skeleton of cellulose behind.

On the other hand it is to be remembered, as one of us has
shown,[1] that the " granulose " has lost all the characters of
starch. Furthermore, the acceptance of cellulose in the residue
is based upon its solubility in ammoniacal solution of oxide of
copper, the loss of its property of swelling in hot water, and the
non-appearance of any coloration by treatment with iodine. But,
on the one hand, amylum is itself soluble to a slight degree in
ammoniacal oxide of copper, and, on the other hand, " granulose " is just as little colored by iodine as the here accepted
cellulose, while the swelling property of starch may also be
destroyed by boiling with glycerin and water. There are thus
not sufficient reasons presented for concurring in Nägeli's pro-
position.

Starch containing water, but not that which has been deprived
of the latter, possesses a highly remarkable attractive power for
iodine. It is capable of so combining with it, that the granule,
the mucilage, or the solution of starch thereby assume colorations which correspond to those peculiar to iodine itself in its
different conditions of aggregation and in its solutions. The
blue, violet, or reddish color which amylum presents when it is
brought in contact with iodine was first observed by Colin and
Gaultier de Clanbry[2] in March, 1814; the other shadings in
violet, red, reddish-yellow, yellow, and brown were studied in
1863, and later by C. Nägeli with great thoroughness. These
shadings of color are limited by the varying reciprocal relations
in the amounts of iodine and starch, as also by the presence of
hydriodic acid and other substances.

Slight amounts of very small starch granules as in chlorophyll
(page 102) may be made more distinctly visible by causing them
to swell by means of caustic alkali, subsequently washing the

---

[1] Flückiger, " Stärke und Cellulose," in Archiv der Pharm, 196 (1871),
7-31.

[2] " Annales de Chimie " 90, p. 93.—Stromeyer, at Göttingen, in December, 1814, pointed out the delicacy of this reaction in Gilbert's " Annalen
der Physik," 49, p. 147.

section with acetic acid and then with water, and finally adding iodine solution.

Though the force with which starch appropriates iodine is quite considerable, it can, nevertheless, not be proved that the product is a chemical compound. Even dialysis, as also gentle warming, and even simple exposure to the air, is capable of eliminating the iodine from the compound.

Only cellulose, under certain conditions, shares this behavior of starch to iodine. Beside *lichenin* (see Index), there is to be mentioned here also the *amyloid* of Schleiden, a form of cellulose which is capable of swelling, is colored blue by iodine, and occurs in the cotyledons of many of the Leguminosæ, for instance, in those of *Tamarindus*. Certain membranes of the hyphæ of lichens also assume with iodine solution a blue color.

The amount of amylum, even in such plants and parts of plants as are abundantly provided therewith, must necessarily be subject to great fluctuations when the before-mentioned function of starch as a reserve substance is taken into consideration.

Potatoes and the rhizomes of the *Maranta* (Arrowroot) afford, for instance, from 9 to 26 per cent of amylum with reference to air-dry substance, and *Sarsaparilla* is likewise a good example of the fact that the percentage of starch is very variable. The statements relating to the quantity of starch present in drugs can therefore be of value only under definite conditions.

The size of starch granules is very variable, although, for the same kind, remaining within narrow boundaries. The largest are found mostly in the underground receptacles of reserve substances (*Solanum tuberosum* as much as 90 μ,[1] *Canna lanuginosa* as much as 170 μ), the smallest in the seeds of some species of *Acacia* (about 1 μ).

An approximate representation of the relative dimensions is given by the following figures :

[1] μ or mic. = micromillimeter = $\frac{1}{1000}$ mm. = 0.000001 m. (compare page 113).

AVERAGE MEASUREMENTS.	According to Wiesner, Höhnel, Wagner, Length in μ.	According to Tschirch[1] Length in μ.
Potato,	60	56.0
Wheat (large granules),	26.9–28	33.0
Wheat (small granules),	6.8	6.0
East Indian Arrowroot,	50–60	——
Maranta,	27–54	32–46
Pea, *Bean,*	} 32–79	29–39
Maize,	15–20	13–19
Rice (divided granules),	5	5–6
Oat (divided granules),	4.4	7–8[2]
Rad. Calumbæ,	As much as 90 μ	
Rhiz. Zedoariæ,	" " 70 μ	according to Flückiger.[3]
Tuber Jalapæ,	" " 60 μ	

Another constituent, related in its functions to starch, is inulin,[4] which Valentine Rose, in 1804, first observed as a deposit from the decoction of the root of *Inula Helenium;* Thomson[5] designated it as *inulin.* It occurs chiefly in the roots of plants of more than one year's growth belonging to the family of the Compositæ, and has been detected elsewhere in but few cases.[6] Prantl[7] has obtained, for example, quite a considerable amount of inulin from the roots of the flowering *Campanula rapuncu-*

[1] The average from 100 measurements. Regarding the relations of shape as well as size, see the very detailed statements of König, " Nahrungs- und Genussmittel," II., 403 *et seq.*

[2] I found (as did Wiesner, in opposition to König), the oat granules to be always larger than those of rice (T.). See A. Tschirch, "Stärkemehlanalysen" in Archiv der Pharm., 223 (1885), pp. 521-532.

[3] " Pharmakognosie des Pflanzenreiches," first edition (1867), pp. 237, 177, 251.

[4] Compare, regarding inulin, Sachs, Bot. Zeit., 1864. Holzner, " Flora," 1864, 1866, 1867, and the publications cited below.

[5] "System of Chemistry," IV. (London, 1817, fifth edition), 75 ; also in earlier editions, previous to the year 1811.

[6] That the Australian *Lerp-Manna,* in opposition to former assumptions, contains no inulin, is now definitely established. Wittstein's Vierteljahrsschrift für prakt. Pharm., XVII. (1866), 161, and XVIII. 1.

[7] " Das Inulin," Munich, 1870, 43.

loides L., and Kraus [1] found it also in the families of the Campanulaceæ, Lobeliaceæ, Goodeniaceæ, and Stylidiaceæ, which, from a systematic point of view, are each and all connected with the Compositæ. Inulin has, moreover, been proved by Kraus to occur in the roots of *Ionidium Ipecacuanha*, of the family of Violaceæ.[2] In the family of Compositæ, inulin possesses the function of amylum ;[3] it is distinguished, however, in general from the latter by the following main points.[4]

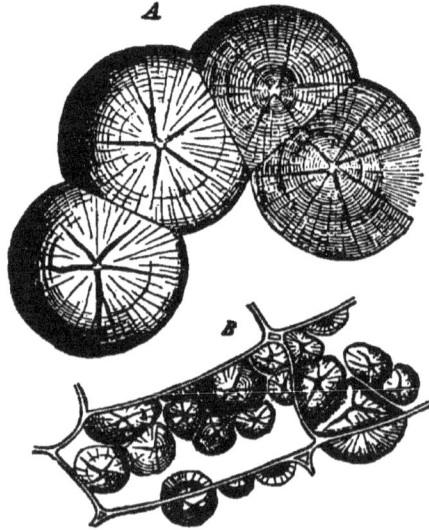

FIG. 54 *a*.—Globular aggregations of crystals (sphæro-crystals) from *Radix Inulæ*, by keeping fresh pieces of the root for a long time in glycerin. *B*, cells filled with Inulin; *A*, separate, strongly magnified aggregations (Sachs). FIG. 52, VI., represents such an aggregation in polarized light (Dippel).

[1] Bot. Zeitung, 1875, 171.
[2] Flückiger, "Pharmakognosie," 1883, 396.
[3] Starch has been found in but a few roots of the Compositæ. Vogl, "Kommentar zur österreich. Pharmakopöe," 1869, p. 347 and Dippel, "Das Mikroskop," II. (1869), 27. Nevertheless, according to Kraus, the chlorophyll granules, the stomata cells, as also the starch-sheaths and sieve-tubes of plants which form inulin contain starch throughout.
[4] Compare further: Dragendorff, "Material zu einer Monographie

In living roots or leaves, inulin does not separate out in a
solid form ; it is only when water is abstracted from the solution, in which it is there contained, that it forms either glass-
like, amorphous masses, or fine, soft, needle-like crystals of the
rhombic system.[1]   The latter may combine to form larger,
radiated, spherical aggregates or *sphæro-crystals* (Fig. 54 *a* and
b), which are best obtained when entire *Dahlia* tubers are placed
in absolute alcohol or concentrated glycerin.   After some days,

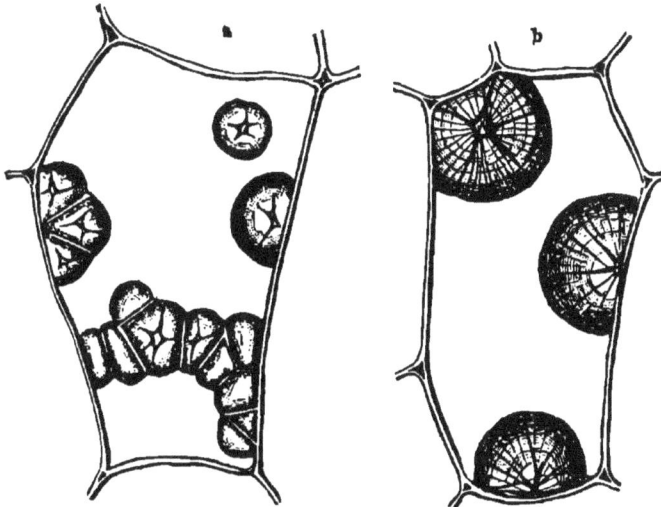

FIG. 54 *b.*—Sphæro-crystals of inulin from *Dahlia* tubers.

in consequence of the slow abstraction of water, the inulin
crystallizes in aggregations, which cannot be obtained by simple
drying.   The leaves of the Compositæ must be prepared for
dehydration by previously boiling them with caustic potassa.

Crystallized inulin, when observed in polarized light, is seen
to be doubly refractive (Fig. 52, p. 119, No. VI.), though less
strikingly so than amylum ; the crossed arms do not appear

des Inulins," Petersburg, 1870. Kiliani, Liebig's Annalen, 205 (1880)
145–190.

[1] Bot. Zeitung, 1876, 368.

very distinctly on the sphæro-crystals, and the amorphous masses are neither doubly refractive nor stratified.

With this deficiency of organic structure is connected also a lesser capacity of combining with water. In opposition to amylum, the composition of which corresponds to the formula $(C_8H_{10}O_5)_2 + 3H_2O$ (= 14.2 per cent of water), air-dry inulin contains only from 5 to 10 per cent of water. On the other hand, it dissolves readily in hot water and separates therefrom unchanged upon cooling, provided the solution had not been exposed for a long time to a higher temperature. In the latter case, the inulin very readily passes into uncrystallizable, lævogyrate sugar.

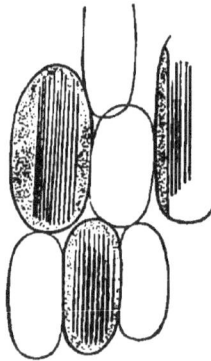

Fig. 55.—Groups of fine needle-shaped crystals (rhaphides) from *Radix Sarsaparillæ*.

The solution of inulin itself likewise deviates the plane of polarization of a ray of light to the left; solutions of starch, which are obtained by the aid of chloral hydrate or by certain salts (page 110), rotate to the right, as does also the crystallizable grape sugar obtained from starch. The aqueous solution of inulin is never paste-like; it is an actual solution in the ordinary sense, while the paste of starch is produced only through a swelling of the granules.

Inulin is not colored by iodine. Indeed, we possess no reagent for it, and are only capable of recognizing it by confirming several of its physical properties.

The amount of inulin contained in the Compositæ is very

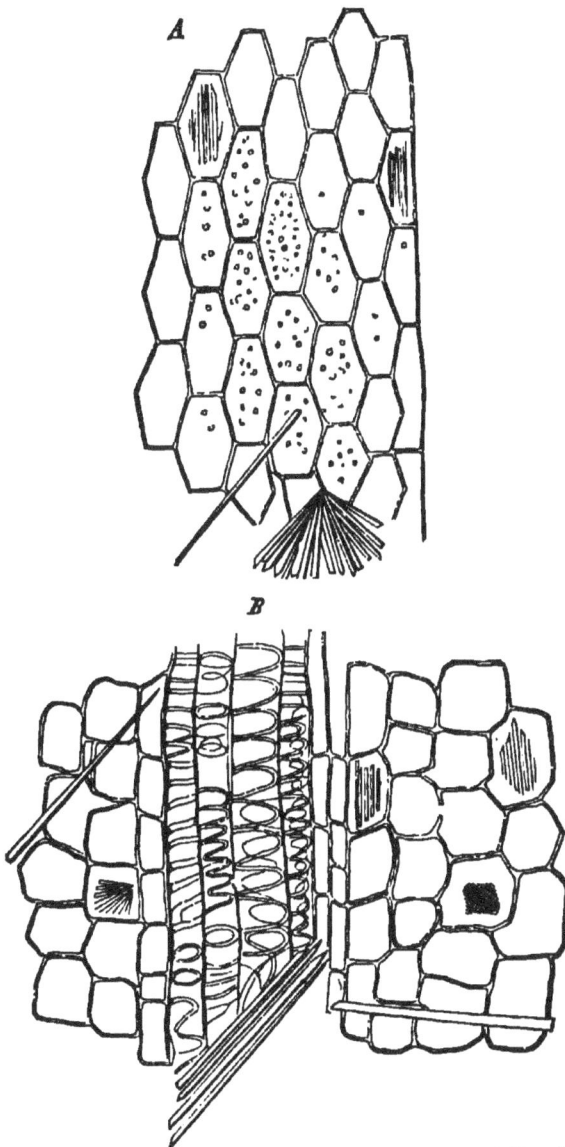

FIG. 56.—*A*, Transverse section; *B*, Longitudinal section from *Bulbus Scillæ*, with numerous prisms of calcium oxalate, which are often approximately 1 mm. in length.

variable, in many cases very slight, as for instance in *Rhizoma Arnicæ*. From dried *Radix Inulæ*, on the other hand, Dragendorff obtained 44 per cent of inulin ; from the root of *Taraxacum*, gathered at Dorpat, Russia, in October, and dried at 100° C., 24.3 per cent, while the same root in March afforded only 1.7 per cent of inulin.

The great periodical fluctuations, and the want of a reagent, may explain why it has not yet been possible to detect inulin in many roots of perennial Compositæ.

Although inulin never occurs in crystals in the living and dried plant, there are other crystalline substances which occur not infrequently in the tissue of the cell. **Calcium oxalate** especially is widely distributed.

In very many plants calcium is deposited in the cells in the form of distinctly crystallized oxalate. This salt mostly corresponds to the formula $CaC_2O_4 + H_2O$, and belongs to the monoclinic (clino-rhombic) system of crystals. Occasionally, however, forms of the quadratic or tetragonal system are also to be seen ; this variety of oxalate contains $3H_2O$.

When calcium oxalate is prepared artificially, and the salt separates rapidly, the first-mentioned compound is obtained either as an indistinct crystalline precipitate, or in well recognizable monoclinic forms ; the quadratic oxalate, on the contrary, crystallizes during the slow evaporation of a hydrochloric acid solution, or also upon the admixture of very little calcium chloride with an extremely dilute solution of oxalic acid.[1] Frequently, under slightly changed conditions, a mixture of both compounds is produced.

The two forms of calcium oxalate are of exceedingly frequent occurrence in the vegetable kingdom. The needle-shaped crystals, *rhaphides*[2] (Fig. 55), appear to belong to the monoclinic

---

[1] With regard to the more precise conditions, compare Souchay and Lenssen, Annalen der Chemie u. Pharm., 100 (1856), 311-325.

[2] From ῥαφίς, the needle. A. de Candolle, in 1826, introduced the term *raphides*, in order to avoid the use of the word crystals, as he supposed (erroneously) those deposits of oxalate of calcium not to consist of crystals.

9

system; these occur separately or in groups, particularly in the
root formations of monocotyledons, very notably in *Bulbus
Scillæ* (Fig. 56), and in *Radix Sarsaparillæ* (Fig. 123 [1]),but also
in stems and leaves, as, for instance, in the *Aloë* (Fig. 63, *cr*).
The undeveloped, crystalline, powdery oxalate, which is met
with, for instance, in the *Cinchona barks*, in *Stipes Dulcamaræ*,
and in *Radix Belladonnæ*, should probably also be considered
here. Such deposits become better recognizable when the sections, freed as much as possible from air, are observed in polarized light. More distinctly and variously developed are the

FIG. 57.   FIG. 58.

 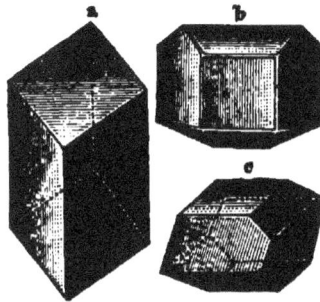

FIG. 57.—Fundamental form of the monoclinic crystals of calcium oxalate, with only
one molecule of water. This form, hendyohedron, resembles in appearance a rhom-
bohedron of the hexagonal system, and is therefore often designated as " rhombohedron-like oxalate."

FIG. 58.—*a*, Hendyohedron ; *b* and *c*, crystals of the monoclinic system in *Cortex
Frangulæ*, derived by truncation from the fundamental form (from Dippel).

crystals which in form approach that of the hendyohedron
(Fig. 57), which may be regarded as the fundamental form of
the monoclinic salt. Very handsome and very regularly developed crystals of this kind occur in *Radix Calumbæ*, *Folia
Hyoscyami* and *Cortex Frangulæ* (Fig. 58), and particularly
also, in a considerable variety of forms, in the non-official
bark of *Liquidambar orientalis* Miller, which yields the *Styrax
liquidus*. In *Cortex Aurantiorum* the crystals are likewise
quite large and are inclined to be sharpened in a striking manner.

[1] Compare also Schleiden, Archiv der Pharm., 1847, Plate I., Fig. 5.

Forms of peculiar appearance, produced by hemitropy, and recognizable by their inwardly inclined angles (Fig. 59), occur in the bark of *Guaiacum officinale* and *Quillaia Saponaria.*[1] Much less widely extended, at least within the sphere of drugs, are well-devoloped forms of the quadratic system (Fig. 60), such as are found for instance in *oak-galls* (Fig. 61). Oxalate crystals of this system also occur in many leaf-stalks, and are particularly handsome in the Cacteæ, in species of *Begonia* and in *Paulownia imperialis* Siebold, furthermore in *Urceolaria scruposa* Ach. and other lichens.

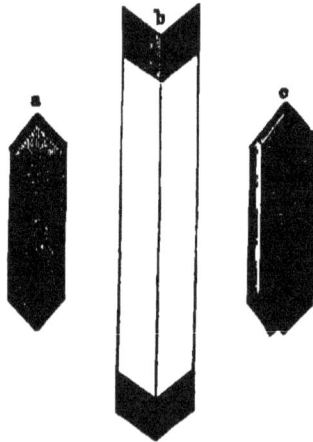

Fig. 59.—Twin crystals of calcium oxalate from *Cortex Guaiaci* or *Cortex Quillaiæ Saponariæ;* a, lying on the lateral surface; c, somewhat turned; b, more strongly magnified and turned to the extent of 90° (Dippel).

In *Rhiz. Rhei, Rad. Saponariæ, Rad. Althææ*, in *Cortex Granati Radicis*, in *figs, cloves*, and in a very large number of other parts of plants belonging to our department (*Fol. Eucalypti*, Figs. 127, 128), the oxalate crystals are most densely crowded together in the form of clusters (Fig. 62); each of which generally occupies a single cell. In these cases only the points of

[1] Further details are given by Holzner, " Krystalle in den Pflanzenzellen," Flora, 1867, 499. Sachs, " Lehrbuch der Botanik" (IV.), 66. Compare also Figs. 102 and 150.

the individual crystals just project, but the true crystallographical shape of the latter has not yet been determined with certitude. The fact that in *Cortex Cascarillæ*, in *Cortex Frangulæ*, in the outer surface of *Fungus Laricis*, in the above-mentioned *Styrax* bark, and in other cases they are accompanied by distinctly recognizable monoclinic crystals of oxalate, argues possi-

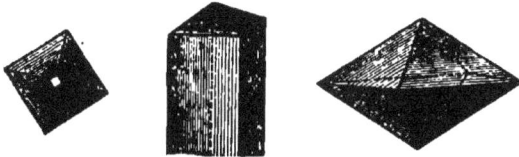

Fig. 60.—Fundamental forms of calcium oxalate crystallizing in the quadratic system with three molecules of water of crystallization.

bly for the assumption that these aggregates or *rosettes* also belong to this system, although in the above-mentioned leaf-

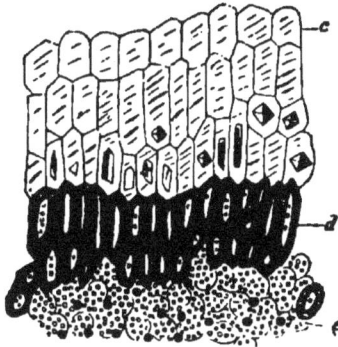

Fig. 61.—Transverse section from an ordinary (Aleppo) oak-gall; *d*, sclerenchymatous layer in the centre ; *c*, tissue outside of and in proximity to this layer, filled with quadratic crystals of oxalate ; *e*, tissue in the interior of the chamber formed by the sclerenchyma, which contains starch and resin.

stalks all transitional forms may also be observed, from the quadratic octahedron to imperfectly developed rosette-shaped aggregates of crystals. Hence it is probable that the oxalate crystallizing in rosettes sometimes belongs to the quadratic and sometimes to the monoclinic system.

The proof that the plant formations in question are really calcium oxalate is readily afforded. The crystals are not soluble in acetic or oxalic acids, but soluble, and without effervescence, in hydrochloric acid; this solution gives, upon the addition of potassium acetate, an abundant precipitate of indistinctly crystallized calcium oxalate.

After short contact with concentrated sulphuric acid, the oxalate crystals are converted into long, lance-shaped crystals of gypsum.

The oxalate crystals are presumably formed in the plant by the gradual confluence of dilute solutions of oxalates with calcium salts. In many cases this occurs with the co-operation of organized structures. The rosettes often inclose an uncrystallized nucleus, and the needle-like tufts of oxalate are fixed, for in-

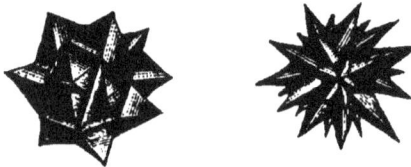

FIG. 62.—Rosettes of calcium oxalate from *Rhubarb* (see Figs. 142, 146) and *Radix Saponariæ*.

stance in *Sarsaparilla* and many other cases, in a mucilaginous (plasmatic) integument; if the oxalate be dissolved in hydrochloric acid (spec. grav. 1.1), the integument remains behind, and may easily be made recognizable by staining, for example, by means of carmine or aniline-red. This covering of protoplasm may be detected also with greater distinctness in *Bulbus Scillæ*. If a fine section of this is moistened with alcohol, a contraction of the mucilaginous contents of the cell ensues, in the middle of which darker granules will appear, which are seen to be crystalline in polarized light. Water dissolves the mucilage and leaves the crystals behind, which, without doubt, are to be regarded as the first rudiments of the oxalate prisms often so handsomely developed in the *Squill*. The latter are surrounded by a sac, and frequently become enlarged to such a degree as to extend through several cells, after their transverse walls are destroyed˙ These crystals often attain nearly 1 mm, in length, so that they become visible even to the unaided eye. The latter character applies also to the imperfectly developed, rhombohedron-like crystals in the wood-parenchyma of *Lignum Sandali rubrum,* the axes which are scarcely less than ½ mm.

According to Emmerling [1] it would appear probable that crystals of calcium oxalate are also formed in the plant through the action of free oxalic acid upon calcium nitrate.

In the cases here referred to, calcium oxalate always occurs as one of the contents of the cell; it has been shown, however, by Count Solms-Laubach [2] that these crystals may also be deposited in the cell-wall itself, especially in the outer wall of epidermis cells.

In regard to the amount of oxalate, the microscopical estimate may lead to inaccurate statements. *Bulbus Scillæ* is apparently rich therein, and nevertheless,a direct estimation of the oxalic acid afforded but 3 per cent of oxalate; in a good *rhubarb* [3] one of us found 7.3 per cent. The greatest abundance of oxalate in the domain of pharmacognosy is presented perhaps by *guaiac* bark, nearly 20.7 per cent. Some lichens are likewise characterized by a large amount of oxalate; thus *Lecanora esculenta* Eversmann, contains 22.8 per cent.

The oxalate crystals are deposits which remain withdrawn from the sphere of vital action (secretions); in the cells which contain them, as a rule, no further developments take place.

Other crystalline compounds of inorganic bases are of exceedingly rare occurrence in plant tissues. Calcium phosphate,

[1] " Berichte der Deutschen Chemisch. Gesellsch.," 1872, p. 782.

[2] Botanische Zeitung, 29 (1871), 458.   Plate VI.— Also in Sachs, " Lehrbuch der Botanik," 1874, 68.

[3] One of the numerous important observations of the eminent apothecary Scheele, who also discovered oxalic acid, relates to the crystals of *rhubarb,* which, in 1782, he recognized as calcium oxalate.   Anton van Leeuwenhoek (1716) had, indeed, previously seen the oxalate of the *sarsaparilla root* and of *orris root.*—Flückiger, " Pharmakognosie," 2d edition, 226, 315, 373.

$CaHPO_4 + 2H_2O$, is found abundantly in a crystalline form [1] in the Indian *Teak-wood* (*Tectona grandis* L., Nat. Ord. Verbenaceæ). Calcium carbonate, which is contained in some plasmodiums, in the cell-membranes of many marine algæ, and in cystoliths (*Ficus, Cannabis, Humulus*[2]), does not show distinctly crystalline forms,[3] or is manifestly amorphous. Crystals of gypsum appear not to be present in plants; since they are soluble in 400 parts of water, the conditions are probably wanting for their formation and maintenance.

Crystals of organic compounds, which are met with in the tissues of drugs, are, however, no rarity. Thus, *asparagin, cubebin, hesperidin, picrotoxin, theobromine,* and *piperine,* which, however, may be presumed to first crystallize during the process of drying the respective drugs. Furthermore, crystallized fats, probably for the most part palmitin and stearin, which are found in many seeds, as, for instance, in the *nutmeg,* in *Cocculus Indicus,* etc. Finally, vanillin in the parenchyma and upon the outer surface of the *Vanilla* (Fig. 83). The crystals which become visible in *Cinchona barks,* after warming their sections in caustic alkali, first appear as a result of this treatment. By very long preservation in glycerin of sections of tissues rich in tannin, crystals of *gallic acid* also occasionally appear, which were not originally present. After a very long preservation of the respective sections, one may also observe the gradual crystallization of amygdalin, filicic acid, and strychnine.

Small granules are frequently found deposited in cells, which acquire with ferric chloride in aqueous, or often better in alcoholic solution, a blue or greenish coloration, so that we may consider them as **tannin,** or as tannin-like formations. On the other hand, they also often become colored blue by iodine, as if they

[1] Kopp-Will's Jahresbericht der Chemie, 1860, p. 531, and 1879, p. 937; "Berichte der Deutschen Chemisch. Gesellschaft," 1877, p. 2,234.—Compare further Just's Bot. Jahresbericht, 1881, I., 402, Reference No. 75.

[2] Flückiger, "Pharmakognosie," 1883, 710; Sachs, "Lehrbuch der Bot.," 1874, 70; Kny, "Botan. Wandtafeln."

[3] Thus also the aggregates in *Castoreum;* compare Flückiger, Grundriss der Pharmakognosie," 1883, 237; further Just's Bot. Jahresbericht, 1881, I., 402, 403.

inclosed starch or had originated therefrom, as indeed both
appear in the same tissues simultaneously, or still more often
alternately. Nevertheless, tannin does not exist to any considerable amount in seeds. The amount of tannin contained in
certain organs, such as barks and fruits, is subject to considerable periodical fluctuations.[1]

Tannin which is deposited in the purest form dissolves when
subjected to examination under water. In order to bring it to
view, the sections must therefore be observed under benzol,
volatile or fatty oils, or other liquids which do not dissolve the
tannic matter; even glycerin suffices, since it dissolves but little
tannin when concentrated. Thus in galls, shapeless masses are
found which almost completely fill the cells.[2] The tannic matter also very frequently penetrates the cell-membranes, so that
the walls of entire tissues become colored after being moistened
with a solution of iron, thus, for instance, the parenchyma of
the *Cinchona barks*, the fibro-vascular bundles and the surroundings of the oil-cells in *cloves*, etc. Thick, hard cell-walls, which
do not become thoroughly moistened by an aqueous solution of
iron, often assume, nevertheless, the blue or green coloration
upon the simultaneous addition of alcohol. These reactions,
however, are perhaps more often produced by derivatives, decomposition products of the tannins, or bodies otherwise related to
them, such as ellagic acid or gallic acid, the presence of which
in nature can, moreover, not yet be accepted with complete certainty. Morin and morin-tannic acid, which react in the same
manner with iron salts, have also not yet been met with in those
parts of plants to which we here devote attention. Furthermore, *pyrocatechin, quercitrin,* and *rutin* must not be omitted
here, which likewise color solutions of ferric salts green. The
first-mentioned substance can, indeed, be cited here only as a
very subordinate constituent of kino, and quercitrin is contained

---

[1] Compare Wiegand, "Sätze über die physiologische Bedeutung des
Gerbestoffes und der Pflanzenfarben." Botanische Zeitung, 1862, 121;
and Kutscher, "Ueber die Verwendung der Gerbsäure im Stoffwechsel
der Pflanze." Flora, 1883.

[2] Berg's "Atlas," Plate 49, Fig. 136.

in *Flores Rosæ gallicæ*, but the latter substance, as well as pyrocatechin, is undoubtedly widely distributed in the vegetable kingdom, and by more exact investigation will probably be found in many other drugs.

Of very frequent occurrence also, and probably quite general in barks in a definite phase of life, is **phloroglucin,**[1] C_6H_3-$(OH)_3$, belonging to the class of phenols.

Resembling the tannic acid of *galls*, or tannin,[2] there are some other tannic matters, not of the same composition, which produce in solutions of ferric salts a blue-black precipitate, thus, the tannin of *Folia Uvæ ursi*, of *oak-bark*, of the *bark of pomegranate-root*, etc. Many others, however, as the tannic acid of the *Cinchona barks*, of *willow* and *elm barks*, that of *Radix Ratanhiæ Peruvianæ*, of *Rhizoma Filicis*, *Rhiz. Tormentillæ*, of *Coffee*, and also *Catechu*, produce with solutions of ferric chloride or ferric salts a green precipitate, while the tannic acid of *rhubarb* gives a blackish-green. In two varieties of *Ratanhia* (*Krameria*), that from Para and that from Savanilla, the tannic acid forming a green coloration with iron salts is accompanied by a predominating amount of acid producing a blue coloration. For the correct discrimination of these colorations, thin sections of the respective drugs must be moistened with a little solution of ferric chloride of the dilution stated under "Microscopical Reagents," and the slides upon which this reaction is carried out, laid upon a sheet of white paper. The experiment is also performed at the same time with the application of a solution of ferrous sulphate, which permits the colorations to appear gradually, in proportion to their oxidation, but often with a greater degree of purity. A highly remarkable occurrence of a substance which affords a magnificent blue color with ferric chloride as well as with ferrous salts is presented by the large cells of the fleshy portion of the fruit of *Siliqua dulcis*.

Between the tannic matters or tannic acids of the two classes

[1] Compare the statements of Tschirch, in Pringsheim's Jahrb. f. wiss. Bot., 1885, and Poulsen's "Botanical Micro-chemistry."
[2] From the French word *tanner*, to tan, of unknown origin.

above indicated, sharp chemical distinctions exist, which are rendered evident, especially, upon dry distillation. When subjected to this treatment, the tannic matters which produce a blue color with ferric salts afford pyrogallol (pyrogallic acid), while those producing a green color with ferric salts, on the contrary, afford pyrocatechin. If the tannic matters are melted with caustic potassa, those giving a blue color with iron salts afford pyrogallol, as in the former case; the other tannic matters, on the other hand, produce protocatechuic acid.

The knowledge of the different members of the chemical family of tannic matters, in their details, is still very fragmentary. A method is also still wanting which meets all demands for the quantitative estimation of tannic acids in all-the numerous cases where they cannot be extracted with a tolerable degree of purity by ether-alcohol, as, for example, from *nutgalls*. Besides, if it be considered that the amount of tannic matter is subject to the fluctuations of vegetation, it cannot be greatly wondered at that the analytical statements relating to it deviate widely from each other. Many such estimations have been made from a technical standpoint, as in the case of *oak-bark*, so that the literature on this subject is quite extensive.[1] The oak-bark appears to be capable of containing a maximum of twenty per cent of tannic matter, or more than any other part of a plant which concerns us here,[2] unless we take into consideration the galls (see the chapter at the end of this work: Pathological Formations). The latter, namely, are to be regarded simply as a morbid accumulation of tannic acid. The gallo-

[1] It may suffice to mention here the following: " Bericht über die Verhandlungen der Commission zur Festellung einer einheitlichen Methode der Gerbstoffbestimmung, geführt am 10. November 1883 zu Berlin. Redaction und Einleitung über die bisherigen Verfahren der quant. Bestimmung des Gerbstoffs von C. Councler. Nebst Untersuchung über die Löwenthal'sche Methode von J. V. Schroeder," large 8vo (IV., pp. 79), Cassel, Fischer, 1885.

[2] The bark of the Australian *Eucalyptus corymbosa* is stated to contain twenty-seven per cent of tannic acid (Jahresbericht der Chemie, 1868, 807), the *Myrobalans* forty-five per cent; *Dividivi*, the pods of *Cæsalpinia coriaria* Willd., fifty-five per cent,

tannic acid, which is present in these malformations to the extent of as much as seventy per cent, is remarkable as a specially distinct member of the family of tannic matters; at least the exceptional cases in which it is supposed to have been elsewhere recognized (in the Myrobalans and in sumach) may still be regarded with doubt.

There may be distinguished physiological and pathological tannic matter. The former is produced normally in the vital process of the plant (thus the tannins of barks,[1] such as that of the *oak*, *quebracho*, and the *willow*). The pathological, on the contrary, is first produced in consequence of an external influence (the puncture of an insect, etc.), that is, in the course of a morbid process (galls). Both forms are also chemically and physically different. Skins are only tanned by the physiological tannin (as in the formation of leather).

The contents of the cells which have so far been treated of, if we except inulin, may be regarded as the organized contents. Besides these, however, there appear a number of unorganized bodies in the cells of plants, which are dissolved in the cell-sap, or deposited in the membranes, and which are not amenable to direct microscopical observation.

In the cell-sap there are dissolved, for example, a portion of the *inorganic salts, dextrin, sugar, plant acids*—the cell-sap always has an acid reaction—and *tannic matters*, many *glucosides*[2] and *bitter principles* (*Aloë*, Fig. 63), *coloring matters, amides,* etc.; in the membranes are deposited many alkaloids (*quinine ?*).

The chief solvent of most of these substances, **water**, evaporates to a large extent upon drying the drugs.[3] How considerable its amount may often be is shown in a striking manner by many

[1] Compare in this connection, F. von Hoehnel, " Die Gerberinden, ein monographischer Beitrag zur technischen Rohstofflehre." Berlin, Oppenheim, 1880.

[2] Γλυκύς sweat, and εἶδος likeness.

[3] To this fact, as it appears, is referred the expression *drug*, German *droge*; the u, which is still frequently inserted in the latter word (drogue), is derived from the Romanic languages, which have appropriated the word. Flückiger, " Geschichte des Wortes Droge," in Archiv der Pharm., 219 (1881), 81.

roots. The younger roots of *Belladonna* lose as much as eighty-five per cent of water; *Radix Taraxaci*, seventy-seven per cent, and juicy fruits still more. All parts of plants, however, retain water which we are wont to designate as hygroscopic water, but which by no means exists in the cells in a liquid form. The amount of this varies very considerably according to the nature of the tissues, and presumably also according to their contents.

The *squill*, which is rich in sugar and mucilage, retains fourteen per cent of hygroscopic water, *Radix Gentianæ* sixteen to eighteen per cent, and *saffron* about twelve per cent. If these substances are completely deprived of water in a drying-closet, or in the cold over sulphuric acid, and are again exposed to the ordinary conditions of preservation, they quickly absorb again about the same amount of water. Drugs which do not have a cellular structure likewise contain definite amounts of water; perfectly air-dry *starch, gum arabic*, and *tragacanth*, for example, thirteen to seventeen per cent. Seeds, on the contrary, and especially those provided with a hard testa, are capable of retaining but a few per cent of water.

After the evaporation of the water, dissolved substances are deposited in a solid form, as has already been mentioned when speaking of inulin. Only a limited number of substances insoluble in water are capable of preserving in the dry tissue so liquid a form as to flow in drops. Such are the volatile oils, the boiling point of which lies from 70 to 150 degrees or more above that of water, in consequence of which they evaporate only to a slight extent with the water at ordinary or only slightly elevated temperatures, and are still further retarded in this respect when they contain resins in solution.

It is remarkable that the milky juice of *Jalap* also still possesses, in the dried drug, a liquid form, and indeed the resin prepared therefrom is capable of retaining water very obstinately.

Besides the loss of water, and probably also of a portion of the volatile oil, many plants experience, upon drying, chemical changes, regarding which we are indebted to Schoonbroodt [1] for

[1] Wiggers-Husemann's Jahresbericht, 1869, 9.

some valuable information, and which deserve to be further studied. Drying changes the properties of many drugs. With some, peculiar substances first appear during the process of drying, while others lose certain principles or acquire a different odor (compare also page 15).

The amount of residue which remains upon drying vegetable objects at from 100 to 110° C., until of constant weight, is termed the *dry weight*. Drugs dried at ordinary temperatures (about 15° C.) are called *air-dried*.

While most parenchymatous cells during life contain, besides protoplasm and cell sap, only little or no **air** (bast-cells, vessels and intercellular spaces contain it abundantly), the cells of dry drugs are generally more or less filled with air, since upon drying this takes the place of water. It is evident from the nature of the case that a complete replenishment of the cells with air is not perceptible by direct observation. In cells which are still succulent and vitally active, on the contrary, and in such which are impregnated with liquids for the purpose of examination, as is necessary in making microscopical preparations, the air bubbles escape as dark rings from the liquid, in consequence of the total reflection of the rays of light. With these the beginner in microscopical observation soon becomes sufficiently acquainted, so as not to mistake them for something else. Tissues filled with air (cork, wood) float upon water, notwithstanding the fact that the specific gravity of cellulose and of cork is greater than that of water. Tissues free from air (for instance, the heart-wood of *Guaiac*), or such from which the air is removed, sink in water, as is also the case with thin laminæ of cork, or with *Lycopodium*, as soon as the air has been expelled therefrom by boiling.

Of the dissolved substances contained in the cells the following may yet be considered:

Sugar is a very widely distributed constituent of drugs. Cane-sugar, and the other varieties of sugar, are so abundantly soluble in water, and probably also in most cell-juices, that even after drying they appear but rarely in a crystallized form or otherwise as a solid constituent of the cells. The more spar-

142   PLANT ANATOMY.

ingly soluble milk-sugar, which does not, however, require more
than seven parts of water for solution at the ordinary temperature, has as yet been found but once (1871) in the vegetable
kingdom, in the fruit of the tropical *Achras Sapota* L.

Grape-sugar (Dextrose),[1] deviating to the right, is of most
frequent occurrence in the vegetable kingdom; it occurs, for
instance, in *grapes, figs, pears, cherries,* in *liquorice-root* and in
tamarinds.

Fruit-sugar (Mucilage-sugar, Lævulose [2]), deviating to the
left, is contained in honey, and often mixed with grape-sugar.

Cane-sugar (Beet-root sugar, Saccharose), deviating to the
right, is contained in the *sugar-cane, sorghum,* the *sugar-beets,*
in *carrots,* and in the sap of the *sugar-maple.* By inversion [3] it
passes into a mixture of one molecule of grape-sugar (dextrose)
and one molecule of fruit-sugar (lævulose), the so-called invert-
sugar[4]; the latter is found in fruits and in honey.

Mycose[5] (Fungus-sugar) is found in fungi, for instance, in
ergot.

Melitose[6] is found in the *Manna* obtained from the leaves of
species of *Eucalyptus* (*Australian Manna*).

Grape sugar is detected in the cells micro-chemically by
Trommer's reaction.[7] The sections are placed successively in a
concentrated solution of sulphate of copper in water (they
should be well washed, but not too long), and in dilute caustic
potassa, and boiled in the latter. If sugar be present, there is
formed in the cells a red, granular precipitate of cuprous oxide,
Cu_2O. (The execution of this reaction requires experience.)

As a result of incisions, there are formed in the *Manna-ash*

[1] *Dexter,* right.
[2] *Lævus,* left.
[3] By boiling with dilute acids.
[4] *Invertere,* to invert, for the reason that invert-sugar deviates to the
left, that is, in an opposite direction to cane-sugar.
[5] Μύκος fungus.
[6] Μέλι honey.
[7] Sachs, "Microchem. Reactionsmethoden." Wiener Academie.
Sitzungsberichte, 1859.

crystalline exudations (*Manna*), which contain as much as eighty per cent of a sweet principle, **mannite** $C_6H_8(OII)_6$.

The glucoside **hesperidin**[1] is contained in unripe fruits of the Aurantieæ, dissolved in the cell-sap, especially in the various species of *Citrus* (very abundantly in *Fructus Aurantii imma-*

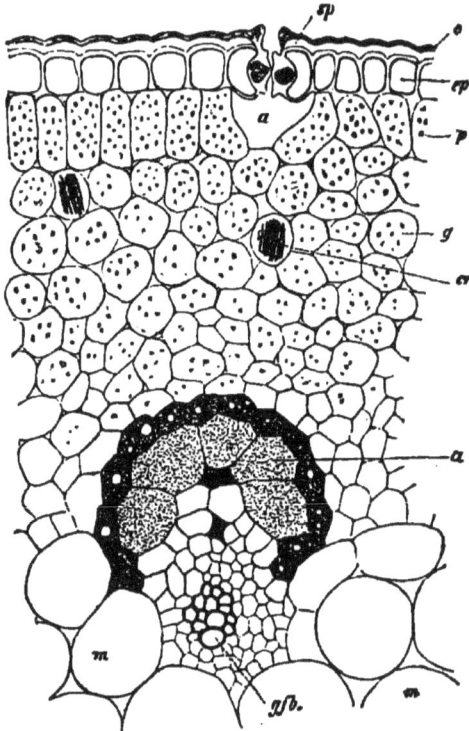

Fig. 68.—Transverse section through the marginal portion of a leaf of *Aloë socotrina*. *ep*, epidermis (*c*, cuticle); *sp*, stoma; *a*, respiratory cavity; *p* and *g*, assimilating tissue; *cr*, crystal cells (with raphides); *a*, cells containing aloes (the large ones contain chromogen); *g/b*, vascular bundle; *m*, medulla containing mucilage.

turi). By immersing the fruits in alcohol, there are produced in the cells sphæro-crystals, similar to those of inulin (Fig. 54),

[1] Pfeffer, Botan. Zeit., 1874, p. 481. Tiemann and Will, "Ber. d. Deutsch. Chem. Ges.," 1881, 946. Virgil named the *Seville oranges* the apples of Hesperides, the daughters of night in Grecian mythology.

which are soluble in slightly alkaline water and in alcohol.
Crystals of this character have also been detected by Adolph
Meyer [1] in the leaves of *Conium maculatum*.

The peculiar bitter substances of *Aloë* leaves, **aloes**, are contained in special cells, which are located directly in front of the
vascular bundles (Fig. 63 *a*), and are confined toward the exterior
by a nucleus-sheath in a single row and with bitter contents. The
before-mentioned cells are short, and occasionally their contents
are crystalline. The entire remaining tissue of *Aloë* leaves contains an abundance of mucilage, but no bitter substances.

The **kino**, from species of *Pterocarpus*, also occurs as a constituent of longitudinally extended cells. [2]

When a part of a plant is incinerated, there remains in the crucible a white residue—the **ash**. Since by careful ignition, in
very many cases, the general outlines of the consumed portion
of the plant remain preserved (leaves of the Gramineæ, hemp
leaves, the shells of diatoms), it follows that the inorganic constituents of the membrane which resist the action of heat (at least
in part) are so finely deposited that the molecule of the membrane can be removed therefrom by incineration, while the direct
connection of the inorganic particles is not destroyed thereby.
Not only in the membrane, however, do we meet with deposits
of mineral constituents, but the contents of the cells are also
abundantly provided therewith. It has already been shown that
protoplasm contains an abundance of salts, that crystals of
inorganic bases occur in the cell-sap, and also that the globoïds
(p. 98) consist of inorganic double salts. The cell-sap, moreover, also contains not inconsiderable amounts of such mineral
constituents of plants as are soluble in water, and these are in
fact the most important, namely, nitrates, phosphates, and
salts of potassium and calcium.

Accordingly, the ash of plants contains all those substances
which are known to, be the necessary nutritive materials, namely: potassium, magnesium, calcium, iron, [3] phosphoric, sul-

[1] Compare also Flückiger, "Pharmakognosie," 1883, p. 663.
[2] See the subsequent chapter on Receptacles for Secretions.
[3] A deficiency of iron is shown in leaves by their becoming yellow
(chlorosis).

phuric and nitric acids, and chlorine. There also occur in it silicium, sodium, manganese, aluminium, iodine, bromine, fluorine, lithium, and other elements.[1]

Plants rich in silicium (grasses, diatomeæ)—they contain it always in the membrane—leave upon incineration a so-called *skeleton of silica*.[2] The Halophytes (salt-plants) especially contain sodium. Manganese is less extended, but is nevertheless found regularly, even though in small amount, for instance, in drugs from the family of Zingiberaceæ.[3] It suffices to reduce to ash a single seed of the *cardamom*, or a still smaller fragment of the fruit-capsule, by heating on the looped end of a platinum wire in the oxidation flame of an ordinary alcohol lamp, and, if necessary, fusing with a little sodium carbonate and a trace of saltpeter, in order to obtain a bead which is colored green by the manganate of the alkali, and which when moistened with acetic acid affords the red permanganate. The same deportment is shown by the root-stocks of this family. The ash of ordinary *cork* (from *Quercus suber*) and that of other species of *cork* is also green from the same cause.

Aluminium is of rare occurrence, but is found in not inconsiderable amounts in the leaves and stems of species of *Lycopodium*.

Iodine and *bromine* occur in the vegetable (and animal)

[1] To interpret the composition of the ash is far more difficult. We are not yet capable of explaining the great differences found therein according to some general law. A very extensive compilation of figures relating to this subject may be found in Wolff's "Aschenanalysen von landwirthschaftlich wichtigen Produkten, Fabrikabfällen und wild wachsenden Pflanzen," Berlin, 1871.

[2] Such skeletons may be prepared by warming small pieces of coarse, firm leaves with concentrated sulphuric or nitric acid and potassium chlorate, expelling the acid, and heating the residue upon platinum-foil (preferably in a current of oxygen) or upon a very thin cover-glass until it becomes white. Tissues which have not previously been treated in this manner often fuse together in consequence of the amount of alkali contained therein.

[3] Flückiger, Pharm. Journ., III. (London, 1872), 208. *Ibid.*, 1886, p. 621; also Amer. Journ. Pharm., 1886, p. 147.

10

inhabitants of the sea; *fluorine* in the testa of the seed of varieties of grain, and *lithium* in *tobacco.*

As *silicium* has in recent times been introduced into organic
compounds in the place of carbon, the supposition is not entirely
unjustifiable that the silicium contained in the cell-wall may be
present in the form of an organic compound.[1]

As is already naturally evident from mechanical principles,
an increased thickening and solidity of the cell-walls corresponds
by no means to a greater amount of incombustible substances.[2]
The delicate tissue, containing air, of peeled *colocynth*, dried at
100° C., afforded 11 per cent of ash, the seed only 2.7 per cent.
Quassia wood from Surinam yields 3.6 per cent, the bark 17.8
per cent of ash; *guaiac wood,* which is so exceptionally dense,
and which consists almost exclusively of strong wood-cells, gives,
nevertheless, scarcely 1 per cent of ash.   Leaves very frequently
contain more than 10 per cent of inorganic constituents, for
instance, *Folia Stramonii* as much as 17, and *tobacco leaves*
occasionally 27 per cent of the substance dried at 100° C.

The developing or merismatic, and the assimilating tissues
(cambium, mesophyll of leaves) and organs (leaves[3] and barks)
are richer in ash than the completely developed and non-assimi-
lating (wood).   The mineral substances migrate from the finished tissues to the places where development is going on.   It
is also seen from this that they must play an important part in
the formative processes of plants.[4]

To obtain the ash in a condition suitable for weighing is often
somewhat difficult, from the fact that many parts of plants,
and especially secreted substances, as gum, resin and sugar,
undergo complete combustion only very gradually.   The incineration of such substances, especially of those rich in nitrogen,

---

[1] "Berichte d. Deutsch. Chem. Ges.," 1872, 568.

[2] That the strength of flexure, for instance, of the grasses, is entirely
independent of the silica contained therein, is shown by water-cultures
in solutions free from silicium.

[3] About the time of the falling of the foliage the leaves become continually poorer in mineral substances.

[4] This is also evident from the experiments with solutions of nutritive
substances.

may be very much accelerated when the objects to be examined are heated on a channeled piece of platinum-foil in a combustion tube in oxygen gas. The same purpose may be attained in a more simple manner, though also more slowly, when the substance which has been carbonized in a platinum capsule is moistened with water, again carefully allowed to dry without decanting the water, and again heated. The water conveys the soluble salts to the unoccupied places of the capsule, and the subsequent admission of air facilitates combustion. If this procedure is repeated several times, a residue free from carbon will in most cases be obtained. Too high a temperature has a retarding effect when salts, such as phosphates of the alkali metals, are present, which fuse together and envelop the carbon; many substances are incinerated more completely by a very moderate degree of heat than at a higher temperature. Very hard shells of seeds offer obstinate resistance to the above procedure of moistening, which may be overcome by triturating the carbonized substance in the capsule, or in the crucible itself, with the aid of a very smooth agate pestle, being careful to avoid loss, and afterwards treating with water. By the strong ignition, which ordinarily is necessary towards the end, carbonic acid is expelled. which must be replaced before the ultimate weighing of the ash, in order to obtain figures which will admit of comparison. This purpose is accomplished by moistening the ash with a little concentrated solution of ammonium carbonate and again drying. It scarcely requires to be mentioned that for reduction to ash the substance employed should previously be dried at 100° C.

The addition of ammonium nitrate or ammonium sulphate also facilitates combustion, especially with substances rich in albumen.

The estimation of the residue left upon combustion is of the greatest practical value, especially for the examination of vegetable powders. For, since every portion of a plant, and thus also every drug, furnishes an amount of ash which fluctuates within definite and often quite narrow limits,[1] the weight of the same

[1] Thus, by way of example, *lycopodium* affords 4, pure *kamala* about

may therefore afford information whether an adulteration with other vegetable or even inorganic powders has taken place.

The estimation of ash must, of course, always be preceded by a microscopical analysis of the substance itself.

II. *The Cell-wall.*

The integument of the cell is called the cell-membrane or cell-wall.

The cell-wall of young cells is a thin membrane which consists of cellulose, and only at a later period becomes variously changed, either chemically through the deposition of other substances, or morphologically through the insertion of molecules of the same kind. Even the membrane of young cells, however, is not perfectly pure cellulose, since it owes its first formation to the protoplasm, remains for a long time in contact with the nitrogenous substances of the latter, and is penetrated by them. Only cellulose which has been purified by means of chemical solvents corresponds to the formula $C_{18}H_{20}O_{18}$.

In living cells, the wall is in most intimate contact with the protoplasm-sac. To this contact is to be referred the **growth of the cell.**

The growth of the cell takes place in a twofold manner, on the one hand by a change of form, and on the other by a transformation of the chemical nature of the cellulose, which, in the course of development of the cells, is capable of assuming a series of new chemical and physical properties.

The change in form of the cell concerns either chiefly its outline, and in this case may be considered surface-growth, or the development of the cell is specially expressed by a thickening of the wall, so that the growth in thickness determines the appearance of the cell. Although both directions of growth are not sharply to be separated, and are essentially based upon the same processes, they nevertheless deviate widely from each other in their results.

2, *lupulin* about 8, *starch* less than 1, *cacao* about 4, *mustard seed* and *flaxseed* from 4 to 4.5, and *pepper* about 5 per cent of ash.

If through a perfectly uniform deposition of new particles of cellulose the mass of the cell-wall becomes equally enriched all around, but not actually thickened, it is compelled to assume a spherical form, and the cells become isodiametric,[1] as in many young tissues. Their mathematical regularity, however, is altered, as soon as the reception of constructive material takes place more energetically in certain places. The outline of the

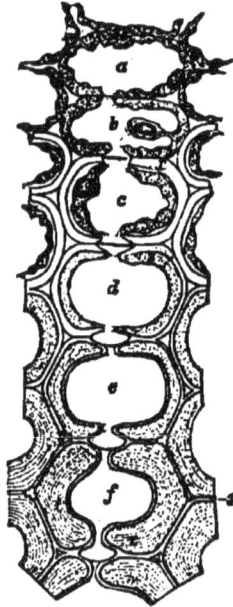

FIG. 64.—Schematic representation of the development of the wall of a wood-cell. *a*, youngest state; *f*, the finished condition. It is only in the first three stages that the nucleus of the cell is preserved (Hartig). A film of intercellular substance closes the pits of the cells.

cells is also very essentially controlled by the fact that they mutually oppose their free expansion. In such cases the form of a sphere becomes flattened to that of a dodecahedron, which is the most uniform of those forms of cells of so frequent occurrence which we designate as spherically-polyhedral, since

Ίσος equal, and διάμητερ diameter.

from their variety and slight regularity they preclude a more
precise definition (Figs. 29, 30, 31, 56, 63, 65, 76).

When the deposition of new cell material does not take place
chiefly in a direction tangential to the cell-wall, but in such a
manner that the latter grows in thickness, this growth can take
place more largely either toward the exterior or toward the
interior. In the first case, prominences of various kinds are
formed (spores and pollen cells, the outer wall of epidermis
cells), in the latter, the cavity of the cell becomes contracted,

FIG. 65.   FIG. 66.

FIG. 65.—Polyhedral parenchyma from *Rhizoma Graminis.*
FIG. 66.—Uncoiling spirals and an annular vessel from *Bulbus Scillæ.*

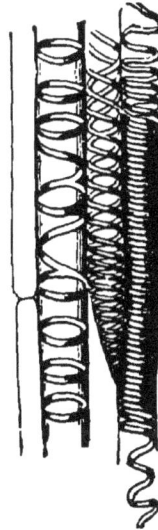

often almost entirely filled up (some bast cells and stone-cells).
A perfectly uniform thickening, however, never takes place, but
the cell-membrane retains in some places its slight thickness.
The appearance of the cells which are subjected to a considerable extent to the growth in thickness is chiefly determined by
the relative extent of the thickened places and those which have
remained thin. If the thickened places are in about the same

proportion as those which have remained thin, the membraue presents, upon a transverse section, a necklace shaped (monili-form) appearance.   Such cells are, for instance, highly charac-teristic of the coffee-bean.[1]   Where the thickened places do not attain great extension, and appear especially upon the inner surface, they often assume the form of rings or spiral bands. Thus originate the spirals in many fibro-vascular bundles, as, for instance, in the *squill* (Fig. 66), as also the net-like and scalariform thickenings (Fig. 68) of the vessels and parenchyma cells (Figs. 67, 83, and 182). ;

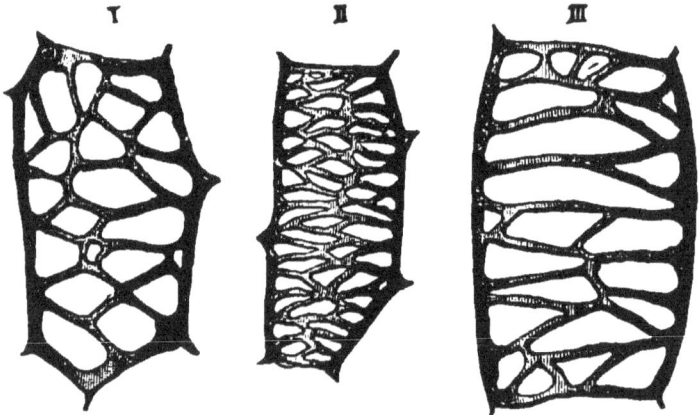

Fig. 67.—Cells with net-shaped thickenings (Dippel). Compare also Fig. 182.

When thickening of the cell-wall extends over the largest part of the inner surface, and exempts but a few expanded dot-shaped places, pores are produced (Fig. 69).   With a considerable thick-ening of the cell-wall such places appear as dots, or by a still greater increase in the thickness of the wall as pore-canals (as in the stone cells, Fig. 70).   Frequently a spiral-shaped arrangement of the dots may be observed (Fig. 71), and the course of the pore-canals also often approaches that of a spiral line.   Many true bast-cells have cleft-shaped dots

[1] Berg's " Atlas," Plate 49, Fig. 131.

arranged in a spiral inclining to the left (compare the subsequent references under Mechanical System of Tissue). A special

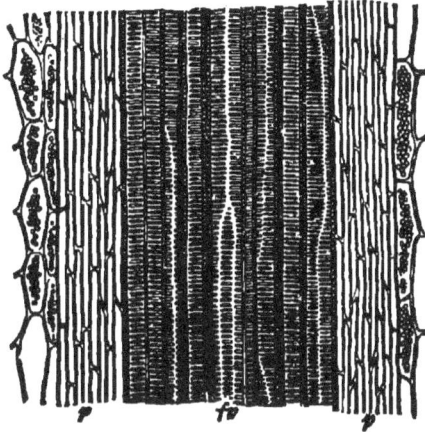

Fig. 68.—Longitudinal section [through vessels with scalariform thickenings (*fv*).— *Rhizoma Filicis* (Berg.)

form of thickening is represented by the bordered pits or areolated dots (Fig. 72). When the cell-wall becomes thickened

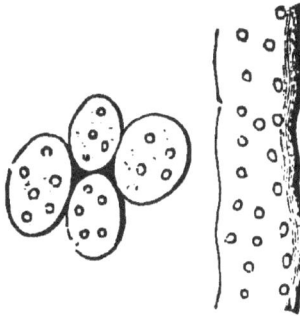

Fig. 69.—Porous cells.

toward the interior around a place which remains thin, a canal will remain open, which in form must approach that of a very obtuse cone, in so far as the walls of the canal are not superposed perpendicularly to the wall of the cell. If in this manner the canal becomes narrower toward the interior, it finally corresponds in form to a somewhat dilated funnel. The upper edge corresponds to the place in the wall which has remained unthickened, and within this circle or border the aperture of the funnel toward the cell-cavity appears as a pit.

Similar areolated pits are wont to appear simultaneously at such places where two cells come in contact by the surface of their walls; the intervening wall, which, moreover, does not always lie in the median line of the pit, but is often, as in

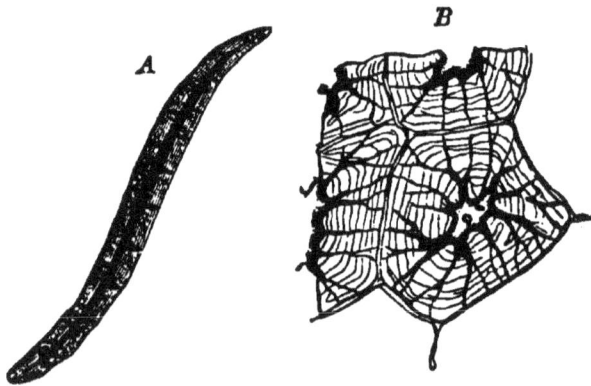

FIG. 70.—Thickened cells with pore-canals. A, Bast-fibre of a *Cinchona Bark*. B, Stone-cells from a nut-shell. (B from Dippel.)

Fig. 64, d, e, f, pressed against one of the apertures (wood-cells of the Coniferæ), disappears by age, so that the space occupied by the pit establishes a direct connection between the two cells (Fig. 72, A, C). These hollow spaces, which sometimes resemble two funnels, one inverted over the other, and which are sometimes arched in a more lens-shaped manner (Fig. 72), are easily recognizable where they occur more isolated. If, however, they are formed in larger number closely beside each other, and if, through increasing thickening, they become gradually contracted in a cleft-like form, more complicated relations are produced,

which are clearly disclosed only in thin and carefully prepared
sections (*Semen Colchici*).

In those cases also where thin places of the cell-membrane
remain preserved only in extremely slight amount and extent,
the growth in thickness does not take place through the simple
deposition of new encircling scales or layers of cellulose.   The

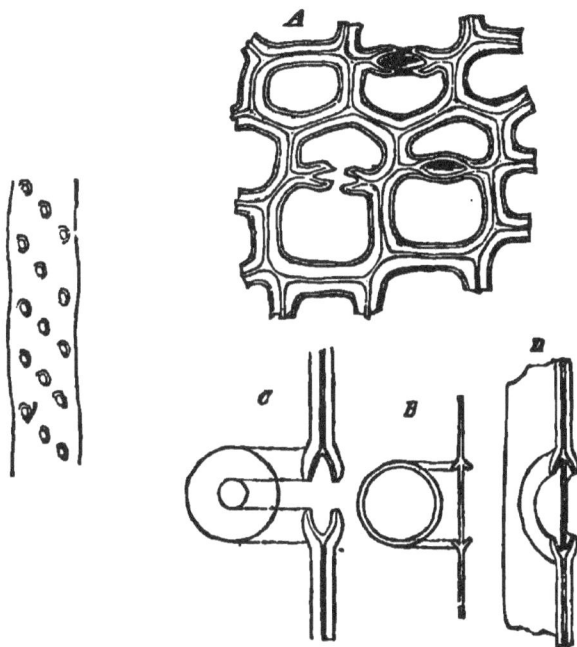

Fig. 71.   Fig. 72.

Fig. 71.—Spirally arranged pits.

Fig. 72.—Areolated dots of the tracheids of fir-wood. *A*, transverse section through
the tracheids or wood-cells, the pits shaded ; *B* and *C*, schematic longitudinal sections;
the spherical lines denoting the circumference of the pit and the border ; *D*, two adjacent pits cut in the direction of length, with the partition-wall still retained (Sachs).

stratification, which is often highly remarkable, depends upon
variations in the amount of contained water and in the conditions of tension of the individual layers ; those containing less
water and which are denser, stand out distinctly in consequence

of their greater capability of refracting the light. Chemical distinctions (various degrees of lignification) also play a part in this. The function of the contained water may be proved by complete desiccation or by more complete swelling, both of which equalize the differences, and often either break up the stratification or obliterate it to a large extent. The bast fibres of the *Cinchona barks* possess almost entirely thickened walls, with distinct stratification (Fig. 73). When they are softened by means of energetic reagents, such as caustic soda, concen

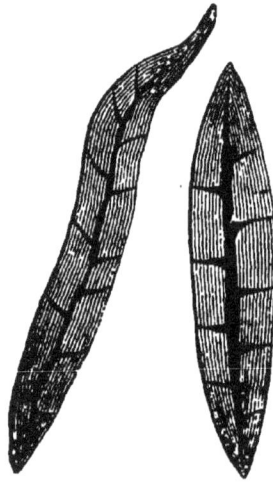

FIG. 73.—Bast-fibres from *Cinchona-barks*.

trated sulphuric acid, or ammoniacal oxide of copper, and the tension of the particles of cellulose becomes equalized, it is distinctly seen that the thickening is not due to a simple concentric succession of layers, but to far more complicated processes.[1] In the *Cinchona* fibres, particularly, there is brought to view in

[1] A more precise elucidation of these remarkable conditions is given by Nägeli, " Bau der vegetabilischen Zellmembran." Sitzungsberichte der Münchener Akademie, June, 1864, page 145. Also Sachs, " Lehrbuch der Botanik," 1873, p. 30 *et seq.* Wiggers and Husemann, Jahresbericht, 1866, 89.

the manner indicated a screw-shaped disposition of the thickening (Fig. 74). Hofmeister[1] found by maceration of these fibres in nitric acid and potassium chlorate, and subsequent pressing, a more scale-shaped arrangement of the layers.

Cells, which have walls of considerable thickness when compared with the diameter of the lumen (cell-cavity), that is such in which the latter is contracted to a very small cleft, are designated as *bast cells* when they are extended in length (Figs. 110 and 111), or as *stone-cells* (Figs. 75 and 76[2]) when they are but

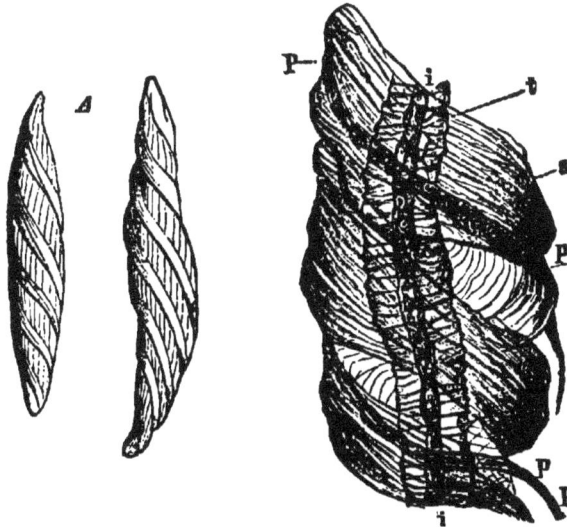

Fig. 74.—*A*, Bast fibres from *Cinchona barks*, boiled with hydrochloric acid.; *P*, the same softened in ammoniacal oxide of copper after treatment with hydrochloric acid, (*P* from Dippel) ; *i*, original size of the cell ; *s*, the swollen layers.

short. The latter, particularly, show a very distinct stratification of the membrane.

[1] " Verhandl. d. Sächs. Gesellsch. d. Wissensch.," X., 1858, p. 32.

[2] They may also be called *sclereids* (derived from σκληρός hard) in opposition to the proper mechanical cells, the *stereids* (from στερεός massive). Compare Tschirch, " Beiträge zur Kenntnis des mechanischen Gewebesystems," Pringsheim's Jahrb., 1885, and " Ber. d. deutsch. botan. Ges., III. (1885), No. 2.

The thickening layers build themselves up, over the places which remain thin, in such a manner that the small canals running toward the centre or the axis of the cell often present a sort of star-shaped arrangement (Fig. 76).

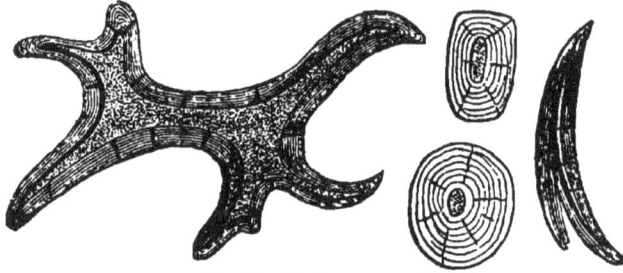

FIG. 75.—Various stone-cells.

The stone-cells or sclereids (see also subsequent references under the section : Mechanical System of Tissue) are widely distributed in many barks, the testa of seeds, seed-vessels, etc. A series of remarkable and manifold forms of them is readily

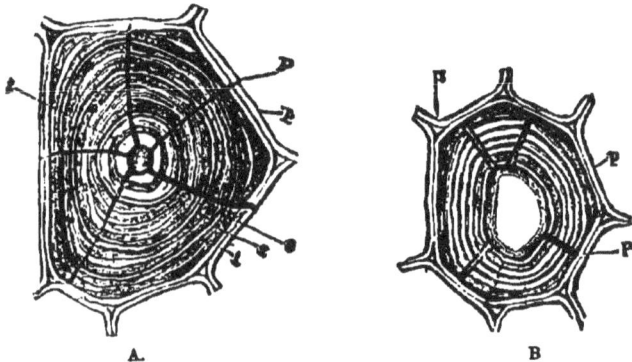

A.   B

FIG. 76.—Stone-cells, whose cavities, *t*, though radiately arranged pore-canals, *p*, are brought into connection with the outer surface, or even with adjacent cells, *i*.   1, 2, 3, thickening layers (Dippel).

afforded, for example, by the *star-anise.* The fruit-stalk contains branched stone-cells,[1] the wall of the capsule such as are nearly cubical.

[1] Vogl, "Nahrungs- und Genussmittel." Vienna, 1872, 111.

The thickening of the cell-wall may also, under certain conditions, confine itself to the corners, thus forming the so-called *collenchyma*,[1] which is met with in the barks and seeds of very many plants.

The relative thickening of individual cells and forms of cells is, moreover, very manifold. While the parenchyma remains mostly thin-walled until the close of life, the wood- and bast-cells become provided with strong walls at a very early period.

Bast-tubes and stone-cells when observed in thin sections under glycerin in polarized light, are seen to be doubly refractive (Fig. 77). A transverse section through cinchona fibres

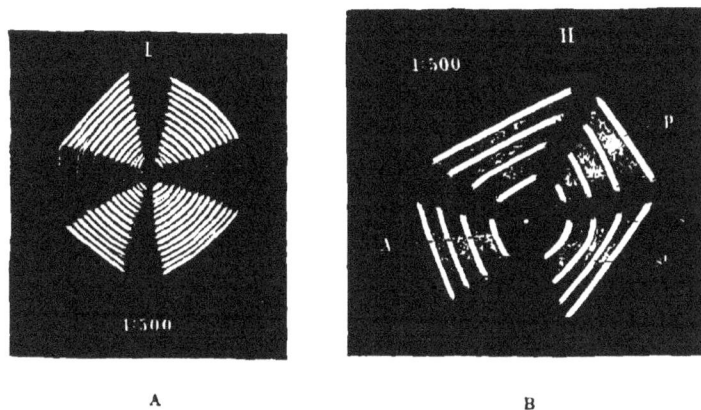

A   B

Fig. 77.—Thin sections through bast-fibres and stone-cells, showing double refraction in polarized light (Dippel), *p, s, s'*, layers of different density.

shows four dark arms of a cross upon a brightly shining ground (Fig. 77. I.).

In the preceding pages, those morphological changes of the cell-wall have been considered which take place in the process of vegetation. It still remains for us to subject the chemico-physical changes to closer consideration.[2]

[1] Derived from κόλλα, glue, since it was formerly, but incorrectly, believed that the collenchyma cells could become mucilaginous.

[2] With regard to the chemistry of the cell-membrane, compare particularly the more recent researches of Cross and Bevan, "The Chemistry of bast-fibres," in the Chem. News, 1882; Webster, "On the analysis of

The cell-wall is subject to chemical and physical changes, either through the deposition of woody matter, *lignin*[1] or cork-fat (*suberin*),[2] as also through a retrograde metamorphosis of the cellulose into gum and mucilage.

All young cell-membranes, and most of the walls of cells, with which we shall become acquainted under the designation of parenchyma, as also many appendages of seeds which are developed as hairs (*Cotton, Asclepias, Eriodendron, Salix*), consist of pure cellulose. The phloëm, *leptom* (sieve-tubes and cambiform tissue), always remains unlignified. Membranes consisting of cellulose show, even by superficial microscopical observation, an entirely different capacity for the refraction of light from lignified and suberified membranes; they appear clearer, more strongly refractive, and jelly-like (collenchyma, the cell-membranes of *Macis*). Cellulose membranes are digestible.

As already intimated on page 123, there are some exceptional cell-walls which, in contact with iodine-water, are colored in a similar manner to amylum. By the treatment of pure cellulose with mineral acids, this faculty may quite generally be imparted to it. The respective sections or objects (for instance, *cotton*) are moistened for an instant with sulphuric acid of the specific gravity 1.84, washed without delay with much water, and then powdered iodine strewn upon the moist preparation, or it is impregnated with iodine-water (see Micro-chemical Reagents). The reaction succeeds with still greater certainty with phosphoric acid, which is first concentrated as much as possible on a water-bath. When hydriodic acid has been formed in an iodine solution which has been long preserved (see Micro-chemical Reagents), such a solution can effect the blue coloration of cellulose without the co-operation of other acids. The reaction admits of demonstration, without further preparation, with moistened parchment-paper, which is sprinkled over with finely-powdered

certain vegetable fibres," *Ibid.*, 1882; Schuppe, " Beiträge zur Chemie des Holzgewebes." Inaugural Dissertation, Dorpat, 1882, in which the older literature is also to be found.

[1] *Lignum*, wood.
[2] *Suber*, cork.

iodine. A solution of chloride of zinc with iodine colors cellulose membranes violet.

The cellulose of fungi, and suberified and lignified membranes are not colored by the above-described treatment; in the case of the latter two, however, this may be brought about if they are previously boiled with nitric acid (specific gravity 1.185) with the occasional addition of a few crystals of potassium chlorate ("Schultze's maceration"). The cellulose of fungi, however, even after this treatment, is not colored by iodine.

Concentrated sulphuric acid alone (specific gravity 1.84) dissolves cellulose with complete chemical change. This is not the case when cellulose is dissolved in ammoniacal oxide of copper (see Micro-chemical Reagents). In contradistinction to the lignified membrane, pure cellulose possesses an exceedingly slight inclination to take up aniline colors (page 161).

Pure cellulose may be prepared by the successive treatment of tissues consisting of this substance (*cotton*, the pith of the *elder* and of *Aralia papyrifera*) by means of caustic potassa, acids, water, alcohol, and ether, or by the precipitation of its solution in ammoniacal oxide of copper by means of water.

One of the most widely distributed modifications of cellulose is formed through the deposition of **lignin** (xylogen [1]). A membrane thus altered is termed lignified.[2]

Lignification appears at very early stages in the so-called wood-cells. The wood-cells which, in dicotyledons, are separated toward the interior by the growth in thickness, already possess lignified membranes long before they become thick. The bast-cells and many stone-cells (sclereïds) are also often lignified. A

[1] Ξύλον wood, and γεννάω produce.

[2] Compare in this connection Stackmann, "Studien über die Zusammensetzung des Holzes," Inaugural Dissertation, Dorpat, 1878, and the previously mentioned dissertation of Schuppe.—M. Niggl, "Ueber die Verholzung der Pflanzenmenbranen" (an historical survey). Jahresbericht der Pollichia, Kaiserslautern, 1881.—Ebermaier, "Physiologische Chemie der Pflanzen," 1882. In this work (p. 175) will also be found statements relating to the amount of lignin contained in some woods. Thus (according to Schulze) *oak-wood* contains 54.12 per cent, and *fir-wood* 41.99 per cent of lignin.

regular lignification of the membrane is also found in the walls of vessels. Lignified membranes refract the light to a less extent than those consisting of pure cellulose, and mostly appear light-yellow under the microscope; they are hard and elastic, and but little capable of swelling.

Lignified membranes are characterized micro-chemically by the fact that with a solution of iodine in chloride of zinc they become yellow (not violet). In ammoniacal oxide of copper and Schultze's macerating liquid (page 160) they do not dissolve. With aniline sulphate and dilute sulphuric acid they become straw-yellow, and with phloroglucin and hydrochloric acid cherry-red; the aniline colors are greedily taken up by them. By boiling with Schultze's mixture or with alkalies, the lignin is removed, and the membranes thus treated then show the cellulose reaction. Morphological alteration, however, by no means goes hand in hand with a change of physical and chemical character. While, for example, the wood-cells, even in quite a young condition when their walls are but very little thickened, are strongly lignified, and the very thin-walled cork-cells are always suberified, the strongly thickened collenchyma, and many bast-cells which are thickened so as to cause the lumen or cavity to disappear, remain unlignified.

The third modification of cellulose is **cork.** This is formed by the deposition between the cellulose molecules of *suberin,* which latter consists for the most part of the glycerin (propenylic) esters of stearic acid and of phellonic [1] acid, $C_{20}H_{42}O_{4}$.[2] Suberin appears to be identical with *cutin.*[3] The epidermis-cells of the more delicate organs of all land plants are covered by a delicate film termed the *cuticle.* The older organs, especially those of the stem, on the contrary, develop on their outer surface a layer consisting of tabular cork-

[1] Φέλλον cork.
[2] Definitely established by Kügler at least for the cork of *Quercus Suber* ("Ber. d. deutsch. botan. Ges.," I., p. xxx., and Inaugural Dissertation, Strassburg, 1884).
[3] Beside suberin, there is also found in cork a wax-like body, *cerin.* The cuticle appears to contain more of the latter than the cork.

11

cells. The cuticle and cork are both produced through suberification (the respective deposition of suberin or cutin in the wall of cellulose).

The cuticle may occasionally become very thick (the leaves of *Eucalyptus, Agave, Aloë*) and in the same manner many layers often develop cork (the *Oak* [1]). Since both the cuticle and the cork are but slightly penetrable by aqueous vapor, they serve as a protection for the organs of the plant against too strong evaporation.

Suberized membranes are mostly brown. They are just as little digestible as lignified membranes, but resist putrefaction very energetically, as does also the cuticle.

Micro-chemically, suberized membranes are characterized by the fact that they dissolve neither in concentrated sulphuric acid nor in ammoniacal oxide of copper. The cork and cuticle therefore remain behind when tissues are treated with sulphuric acid.

It is, however, to be observed in this connection that the membranes of the tissues of drugs, in consequence of their strong infiltration with the constituents of cells, which takes place during the process of drying, often resist very obstinately the action of reagents, even when no suberification, etc., has taken place. It is only after repeated boiling with alcohol, water, and ether, that such membranes are made accessible to reagents.

Suberin cannot be removed from the membranes by the ordinary solvent of fats, but only through the action of alcoholic potassa. It is therefore very firmly (perhaps chemically ?) combined with the cellulose.

Closely related to the lignified and suberized membranes is the so-called **intercellular substance** (middle lamella), or that substance which cements the cells to each other (x in Fig. 64).

The intercellular substance [2] is insoluble in concentrated sulphuric acid and in ammoniacal oxide of copper; on the other

[1] Compare also the section : Epidermal Tissue.
[2] Compare herewith, among others, R. F. Solla, " Beiträge zur näheren Kenntniss der chemischen und physikal. Beschaffenheit der Intercellularsubstanz," in Oesterr. botan. Zeitschr., 1879.

hand, it is soluble in nitric acid with the addition of potassium chlorate (though not without decomposition) and in a hot solution of caustic potassa. Aniline colors are strongly absorbed by it.

The cell-membrane of the living plant (and to an increased degree that of drugs) contains, however, beside the deposited substances just mentioned, not only the organic constituents of the cell which enter it by infiltration, but also abundant amounts of inorganic compounds,[1] as has already been explained (page 144 *et seq.*). These are mostly deposited molecularly (as silicium),[2] more rarely in form of crystals (as in the spicula-cells of *Welwitschia mirabilis* and the epidermis of *Dracæna* leaves).

As has previously been mentioned, the cuticle contains not only suberin, but also wax-like bodies.[3] Occasionally **wax** issues from the membrane, and then forms coatings consisting of granules, small staffs, or crusts. If these are distributed in slight amount over the epidermis, the plant organs assume the appearance of being covered with hoar-frost (*pruinosus*), thus the leaves of *Eucalyptus*, *Ricinus* and *Cabbage*, *Plums*, and many other allied fruits, and *Juniper berries*. In some plants, however (especially several *palms*, Anacardiaceæ, Myricaceæ), the secretion is so considerable that the wax may be collected in large amounts, as, for instance, the Carnauba-wax, from the young leaves of the East Brazilian palm, *Copernicia cerifera* Martius.

The different varieties of wax are esters (compound ethers) of the fatty acids; upon saponification they do not, however, afford glycerin, but other (monatomic, not triatomic) alcohols.

The varieties of **gum** and the **mucilages**, being closely related to cellulose, must be considered in connection with the latter.[4]

[1] The amount of mineral constituents of the membranes is very variable. The best quality of cotton, dried at 100° C., affords but 1.13 per cent of ash.

[2] In the siliceous coatings of diatoms, and in the grasses.

[3] Compare De Bary, Botan. Zeit., 1871, Nos. 9, 10, 11, 34; and " Anatomie, "p. 87 et seq.

[4] Compare also Valenta, " Die Klebe- und Verdickungsmittel."—Cassel, 1884.

The relations of these bodies to each other, as also to cellulose, have, however, not yet been made clear.

According to Giraud [1] these substances may be grouped in the following manner:

1. Ordinary varieties of gum: arabin, bassorin, cerasin;
2. Pectose: *gum tragacanth* (adragantin);
3. Plant mucilages in a more restricted sense:
    (a) insoluble in alkalies and dilute acids (the cellulose of quince-mucilage);
    (b) insoluble in alkalies, and forming with acids glucose and a variety of dextrin: *flaxseed*, mucilage of *Irish moss;*
    (c) soluble in hot, concentrated alkalies, and converted by acids into dextrin and glucose.

Mucilaginous substances, in the broadest sense, are also distinguished by their behavior to nitric acid; some afford with it mucic acid, $C_4H_4(OH)_4(COOH)_2$, while others do not. Furthermore, the aqueous solutions of many varieties of gum are precipitated by the normal acetate of lead, while others are only precipitated by the basic acetate. [2]

Gum arabic and *cherry-tree gum* (cerasin), as well as *tragacanth*, are formed by a retrograde metamorphosis of the cell-membrane, [3] that is, by a pathological process; the former two by a conversion of the membranes of the peripheral layers of the "horn-bast prosenchyma" [4] into gum, the latter through a meta-

[1] Compt. rend., 80, 477; compare also Husemann and Hilger, "Die Pflanzenstoffe," I. (1882), 131.

[2] Compare Kirchner and Tollens, Liebig's Annalen, 175 (1874), 205.

[3] In reference to this and the following statements compare: Mohl, Bot. Zeit., 1857, 33.—Frank, Jour. f. Pract. Chem., 95, 479; idem, Pringsheim's Jahrb. für wissenschaftl. Botanik., V., 25.—Wigand, "Ueber die Desorganisation der Pflanzenzelle," in Pringsheim's Jahrb., III., 115.—Prillieux, "La formation de la gomme." Ann. des sc. nat., 6 Ser., Bot. I., 176.

[4] With regard to horn-bast, compare Wigand, Flora 1877, p. 369, and "Lehrbuch der Pharmacognosie," 1879, pp. 9 and 38; further Flückiger, "Pharmakognosie," 349. The word "horn-bast" should be expunged from the modern terminology.

morphosis of the membranes in the medulla and medullary rays (Fig. 79).

In the Amygdaleæ, however, gum also occurs abundantly in the vessels and other elements of the wood, and often even en-

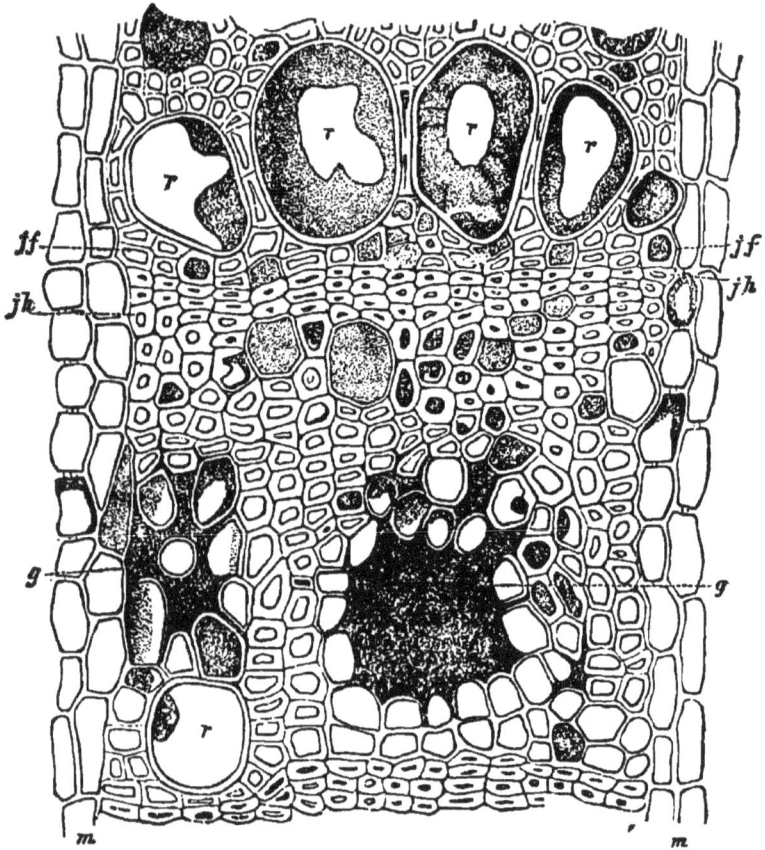

FIG. 78.—The formation of gum in cherry-wood; *g*, aggregates of gum formed through metamorphsis of the cell-membranes; *r*, vessels more or less filled with gum; *m*, medullary rays; *jf*, annual ring, spring wood; *jh*, annual ring, autumn wood (Tschirch).

tire groups of cells of the woody structure suffer a conversion into gum (Fig. 78 *g*), as in the so-called gum disease.

Beijerinck[1] attributes the origin of gum arabic, the "gummosis" of species of *Acacia* of Africa, to the fungus *Pleospora gummipara* Oudemans; another fungus, *Coryneum Beijerinckii* Oudem., causes the gummosis of the Amygdaleæ. Frank does not concur in this view, and Wiesner (*Botan. Zeit.*, 1885, p. 577, also *Ber. d. Deutsch Chem. Ges.*, 1885. Referate p. 639), recently attempted to show that the transformation of cellulose (and starch) into gum or mucilage is due to a peculiar ferment, a "diastatic enzyme." At all events, the true gummosis, which is certainly a pathological process, must be separated from the gum formation which serves as a protection to tissues (see subsequent references), and which is only intelligible from a physiological point of view. They are also distinguished from each other by the fact that the "pathological gum"—and only this is of interest to us here—is formed through a metamorphosis of the membrane, while the "physiological gum" represents an exudation of the membranes into the cell-cavities.

The transformation of cellulose into gum and mucilage can also take place without so great an alteration of the cells and tissues as in the case with *gum arabic* and *tragacanth*. In such a case only one layer of the membrane becomes metamorphosed, as is shown, for instance, by the conversion into mucilage of the filamentous Algæ. The gum mucilage of the glandular hair (colleters) or many foliage buds, which is often mixed with volatile oil and resin, is formed through the conversion into mucilage of a membranous layer (collagen layer[2]) lying underneath the cuticle of the glandular hair. The mucilage of *quince seed* and of *flaxseed* is probably also formed primarily through a conversion into mucilage of only the secondary membrane of the epidermis cells[3] of the respective seeds (Frank). The seeds of many of the Papilionaceæ, for instance, those of *Trigo-*

[1] "Onderzoekingen over de Besmettelijkheid der Gomziekte bij planten." Amsterdam, Joh. Müller, 1884, 4to, 46 pages, 2 plates.

[2] Κέλλα glue, and γεννάω to produce. Hanstein, "Ueber die Organe der Harz- und Schleimabsonderung in den Laubknospen." Bot. Zeit., 1868, No. 43.

[3] Berg's "Atlas,",Plate XLVI., Figs. 122, 123 6.

nella fænum græcum (*fenugreek seed*), present an illustration of the formation of mucilage occurring in the inner tissue, not in the epidermis.[1]

In leaves, mucilage appears to be formed but rarely in larger amounts. A very remarkable example of this character[2] is presented by the *Buchu leaves* from *Barosma crenulata* Hooker and other species.

But mucilages are also formed in the plant, apparently without this direct participation of the membrane. They then fill either all the cells of the tissue (*Irish moss*), often in combination with starch (*Sphærococcus lichenoides*), or are confined to individual cells, which are often distinguished by their shape and size (*cinnamon, elm bark, salep*[3]), or, finally, are given off by intercellular receptacles of secretions (Cycadeæ). Even the gum which exudes upon wounded places, for the purpose of closing the vessels—and which, therefore, serves a physiological purpose—is not formed through a conversion of the membrane into gum, but is secreted by the same in the form of drops[4] (see above).

The designation, *bassorin*, has injudiciously[5] been transferred to a part of the mucilages. Solutions of plant mucilage are not only precipitated by basic acetate (subacetate) of lead, but also by the neutral acetate (sugar of lead). Plant mucilage from its various sources presents, however, in its behavior to water, all gradations, from complete solubility to mere swelling, accompanied by but extremely slight solution. For the purpose of microscopical examination of tissues containing mucilage, those liquids are therefore useful which act to a less extent upon

[1] Flückiger, " Pharmakognosie," 1883, 934.

[2] Flückiger, Schweizerische Wochenschrift für Pharmacie, 1873, p. 435; Flückiger and Hanbury, " Pharmacographia," 1879, p. 109; Radlkofer, "Sapindaceen-Gattung Serjania." Munich, 1875, 100.

[3] Berg's " Atlas," Plate XXIII., Fig. 57.

[4] Frank, " Berichte d. deutsch. botan. Ges.," II. (1884), 822.

[5] Injudicious in so far as under the name of *Bassora Gum* different and not accurately known varieties of mucilage, similar to *tragacanth*, have been grouped together; the expression *bassorin* is, therefore, not capable of precise definition; and should be abandoned.

the latter, such as concentrated glycerin, alcohol, and fatty or volatile oils. The mucilage then appears contracted to a mass which no longer completely fills the cell, as, for instance, in *Bulbus Scillæ*. Occasionally the masses of mucilage show a stratification, which is rendered more prominent upon the addition of alcohol, as may be seen in *Radix Althææ*. In such cases it is to be assumed that a gradual, even though but partial, conversion of the cell-wall into mucilage has taken place, especially when the mucilage, as in *salep*, is colored blue by iodine and sulphuric acid, or is even soluble in ammoniacal oxide of copper, like pure cellulose. The latter is the case with the terminal

FIG. 79.—Transverse section through *tragacanth*, in which may still be seen the remnants of the cell-membrane which has been converted into gum, and also isolated starch granules.

member of the cellulose series, lichenin (see pages 123 and 170), which is related to the varieties of mucilage. That the cell-walls are capable of passing entirely into mucilage has already been noted (page 164). In the formation of *tragacanth,* not only the cell-membranes participate, but also the starch granules which were previously deposited in the tissue, and which to a slight extent are still retained as such in the *tragacanth* (Fig. 79). Other constituents of the cells also occasionally take part in the formation of gum and mucilage.

Considerable differences, which have, however, been demonstrated as yet only by a few examples,[1] are also presented by the various kinds of mucilage from an optical point of view, some of them rotating the plane of polarized light to the left, in the same manner as ordinary gum, while other mucilages rotate it to the right.

With regard to their chemical character, gums and mucilages are but little known, and are with difficulty freed from inorganic constituents and nitrogenous substances.[2] *Gum arabic* appears to be composed of the calcium, potassium, and magnesium salts of arabic acid. If the formula $Ca(C_{12}H_{21}O_{11})_2 + 3H_2O$ is assigned to gum arabic, it must contain 13.3 per cent of water and 1.9 per cent of calcium; these numbers nearly correspond to the actual proportions.

It is also scarcely possible to characterize gums microchemically. They mostly swell in water (not the gum produced by wounds) and are not rendered blue by iodine, or by iodine with sulphuric acid. The *plant mucilages* are colored yellow or blue by iodine, and blue or violet-brownish by iodine with sulphuric acid. Both are insoluble in ammoniacal oxide of copper. The *amyloid* of Schleiden, which should also be considered here, is colored blue by iodine, yellow by iodine water, and is soluble in boiling water. In some cases, for instance, in *Cydonia* and *salep*, the mucilage retains the capability of being colored from reddish to blue by iodine, after treatment with sulphuric acid, and in this respect stands one step nearer to cellulose. It does not follow from this, however, that mucilages always originate from cellulose. In *Semen Cydoniæ, Sem. Lini, Sem. Sinapis albæ*, and also in the seeds of *Plantago Psyllium*, before they ripen, and

[1] Wiggers-Husemann's Jahresbericht, 1869, 154, top.

[2] *Tragacanth* affords three per cent of ash. In mucilage of *Irish moss*, even after repeated purification, there are still contained sixteen per cent of inorganic substances and 0.88 per cent of nitrogen (= six per cent of albumen): Wiggers-Husemann's Jahresbericht der Pharm., 1868, 33. With reference to many other varieties of mucilage compare Frank, Pringsheim's Jahrb. für wissenschaftliche Botanik., V. (1866), 161.

before mucilage makes its appearance in the respective cells,
there are found starch granules, which afterward disappear—a
circumstance which very probably stands in definite relation to
the formation of mucilage.

We shall also meet with a transformation and solution of
cellulose later on, when we come to the consideration the origin
of cell fusions. In the formation of vessels, sieve-tubes and
lacticiferous ducts, namely, a resorption (solution) of the transverse walls consisting of cellular substance takes place. Besides
this process of solution, there also occurs, during the formation
of the lysigenic balsam ducts (see also under " Receptacles for
Secretions "), a transformation of the cellulose into secretions.

Thus, for example, the membrane may become converted into
resin [1] (see Index references to the latter). Such a transformation appears also to be the case in the formation of the resin of
Polyporus officinalis Fries.

With the varieties of gum and mucilages are also connected
the **pectic substances,** the knowledge of which is still very
incomplete.

In close connection with the bodies which have here been
treated of stands **lichenin** [2] (lichen-starch, amylo-cellulose), a
carbo-hydrate deposited in *Cetraria islandica*, in *Usnea, Parme-
lia* and *Cladonia*. According to Berg, the lichenin of the first-
named lichen, the *Iceland moss*, is a mixture of two substances,
one of which is colored blue by iodine and is dissolved by
chloride of zinc and ammoniacal oxide of copper.[3]

---

[1] The formation of resin can take place: 1. As a true secretion through
proper organs of secretion. 2. By the liquefaction of the outer walls of
certain cells. 3. By a metamorphosis of the entire cell-wall and contents of the cell (lysigenic and pathologic receptacles for resin). 4. By
a transformation of certain constituent bodies, increasing the resin
formed according to 2 and 3. (Compare Hanausek, " Jahresbericht der
Handelsschule in Krems," 1880). See also subsequent references under
" Receptacles for Secretions."

[2] From *lichen*.

[3] Jahresbericht der Pharmacie, 1873, 21. Compare Flückiger,
" Pharmakognosie," second edition, 273, and " Ueber Stärke und Cellulose," in Archiv der Pharm., 196 (1871), 27.

II. Forms of Cells.

Notwithstanding the unlimited variety of forms of developed plant cells, they, nevertheless, show very similar outlines in a young condition. In every case where cells can develop unobstructed, they assume a spherical shape (Fig. 80), which is the fundamental form of all cells (*Saccharomyces*, spore cells, pollen, heads of glands, the cells of soft tissue, for instance, of the medulla).

The subsequent distinctions in form are produced either through unequal surface growth, or growth in thickness or length of the cell, or through the pressure of contiguous cells.

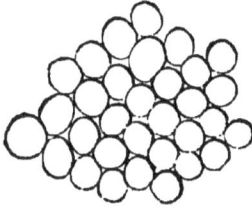

FIG. 80.   FIG. 81.

FIG. 80.—Spheroidal cells. Isodiametric parenchyma.
FIG. 81.—Parenchymatous tissue from the pith of the *Elder*.

If the surface growth does not proceed uniformly, there are formed a great variety of elliptical, tabular or hemispherical, sinuate, star-shaped (Fig. 152) or plaited cells.

If the growth in thickness is unequal, all the forms are developed which have previously been mentioned when considering the growth in thickness of the membrane (page 149): the pitted, scalariform, annular and spirally thickened cells (vessels, Figs. 66 and 68), the stone-cells and bast-cells (sclereïds,[1] Figs. 70, 110, 115, 116, 117). If the growth in thickness is confined to the corners, collenchyma is formed (Fig. 109 *b*).

If the growth in length is unequal, that is, chiefly confined to two opposite sides, there are produced elongated forms of

[1] See page 156, foot note.

cells, bast and wood-cells (Fig. 136), sieve-tubes (Figs. 147 and 148), furthermore sickle-shaped, and S- or U-shaped cells.

Mutual pressure also causes manifold differences of form. Thus from roundish cells (Fig. 80), polyhedral (Fig. 81), and more or less rectilineal forms are produced. It is only in very delicate and soft tissues (the fleshy part of fruits, medulla, leaf-cells), and in those places where the membrane reaches the outer air (outer wall of the epidermis and bordering membranes of the intercellular spaces), that the spherical outlines remain preserved; in firm and hard tissues (wood, Fig. 180, bast groups, Figs. 111, 112, 113) all cells are seen to possess, on a transverse section, more or less flattened forms with a rectilineal border.

There are ordinarily distinguished, according to the scheme first proposed by Link:[1]

1. *Parenchyma*,[2] thin-walled, mostly roundish-polyhedral and isodiametric cells: cells of the fundamental tissue, of the medulla, of the fleshy part of fruits, merenchyma[3] of the leaves, and when extended in a palisade-like manner, palisade-cells[4] (Figs. 80, 81, 85, 108, 127, 128, 129).

2. *Prosenchyma*,[5] consisting of thick-walled, more or less elongated, spindle-shaped cells, with the ends wedged into each other: wood-cells, bast-cells (Figs. 139, 136, 110, 111).

Striking and convenient as the discrimination between prosenchyma and parenchyma appears, yet it does not admit of sharp application.

Fungi and lichens are composed of thread-shaped cells, *hyphæ*,[6] which continue to grow at the ends and mostly divide and branch by transverse walls (Fig. 82). They are not only

[1] "Grundlehren der Anatomie und Physiologie der Pflanzen." Göttingen, 1807.
[2] Παρά beside, thereon, and εγχυμα that which is poured in, the cells considered as standing upon each other.
[3] An expression introduced by Meyen ("Phytonomie," Berlin, 1830).
[4] *Palus*, i, masc., a stake.
[5] Πρός towards, between, and εγχυμα (see above), the cells considered as inserted between each other.
[6] Ὑφή the tissue.

densely interwoven, but also cling together with great tenacity, unless they inclose hollow spaces. The tissue (pseudo-paren-chyma)[1] of sclerotiums, for instance of *Secale cornutum*, consists of remarkably short hyphæ, so that upon thin sections it has the appearance of parenchyma. It is only upon a longitudinal section, softened by a dilute solution of chromic acid (see Micro-chemical Reagents), that the threadlike nature of these hyphæ is likewise clearly brought to view. Notwithstanding their slight length, they are very firmly connected with each other.

Through subsequent resorption of the transverse walls of a more or less elongated row of cells (cell-fusions), long tubes may

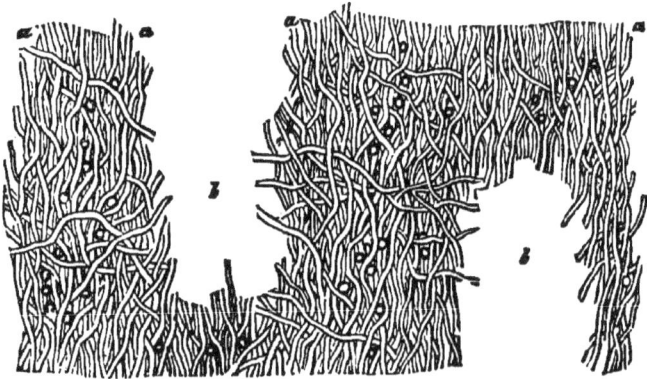

Fig. 82.—Hyphæ from *Fungus Laricis*. *a*, Hyphæ; *b*, longitudinal section through the hollow spaces (pores). (Berg.)

be produced which possess the most varied physiological func-tions, sometimes as vessels (in the woody structure), sometimes as sieve-tubes (in the phloëm), and sometimes as lacticiferous ducts (in the fundamental tissue); all being thus designed for conducting purposes.

If the cells lie densely upon each other, they are firmly cemented together by the *intercellular substance*[2] (middle lamella) (Fig. 64 *x*); if, on the contrary, they are not in contact with each

[1] Ψεῦδος illusion.
[2] *Inter*, between, and *cellula*, cell.

other on all sides, there appear between them (especially at the corners) *intercellular spaces*, which are mostly filled with air (Figs. 127, 129, 151, 152, 155).

III. Cellular Tissue.

With the exception of the one-celled plants and plant organs (*Saccharomyces*, trichomes, *Lycopodium*) and some fungi and algæ represented only by simple cellular threads, all plants consist of cellular tissue,[1] that is, of cells (aggregations of cells) in every form of arrangement. All the cells of such tissue are never completely uniform, but the individual parts become distinguished at an early period in a more or less pronounced degree. While the cell of the algæ performs conjointly all the functions which are required of the plant, in the higher plants a division of the work takes place in such a manner that some forms of tissue undertake one task and others another.

Through this division of work there are then produced in the body of the plant **anatomico-physiological systems of tissue.** Such a system of tissue is, therefore, a union of cells, complete within itself, and connected by their entire physiological deportment.

This differentiation of the body of the plant, however, is not noticeable until the later stages of development. At the places of development, the growing points (apex of the stem, tip of the root), such a difference in the tissues is not yet perceptible. The tissue here consists rather of uniform, more or less isodiametric, thin-walled cells containing protoplasm, and in a state of most active division. Such a tissue is termed **developing tissue,** or *meristem.*[2] As such it stands in opposition to all the remaining tissues, which are also collectively comprehended under the name of **permanent tissue.** The cambium (Fig. 138) is also such a developing tissue.

---

[1] The inner juicy tissue of maturing fruits (*tamarinds, juniper berries, oranges,* stone-fruits) is resolvable into individual cells, but these are always held together by their surroundings.

[2] Μερίζω I divide.

While, namely, the cells of the developing tissue have not yet assumed a definite, permanent form, but become altered by division and mutual dislocation, the cells of the permanent tissue are conclusively defined with regard to their form, or are subsequently but little changed. By far the most drugs consist of permanent tissue.

In the angiosperms there are to be distinguished at the growing point three meristem zones: the *dermatogen*[1] from which is formed the epidermis; the *periblem*,[2] from which is produced the bark; and the *plerom*,[3] from which are formed the vascular bundles and the medulla. In angiospermous roots there is, in addition, the *calyptrogen*,[4] which represents the developing tissue of the root-cap.

IV. Systems of Tissue.

If, in grouping the forms of tissue as systems of tissue, the latter are viewed not alone from a purely anatomico-topographical standpoint, but if, at the same time the question arises, in which manner the various tissues are of equal value physiologically, they may be divided in the following manner:

1. *The epidermal system.* Function: the protection of the organs from without.

2. *The mechanical system.* Function: to give stability to the plant.

3. *The assimilating system.* Function: assimilation of the carbon.

4. *The conducting system.* Function: conduction, especially the conduction of water and nutritive salts from the soil, and to conduct away the products of assimilation.

5. *The storing system.* Function: the storage of reserve nutritive substances and of water.

6. *The aërating system.* Function: the aëration of the organs.

[1] $\Delta\acute{\epsilon}\rho\mu\alpha$ skin, and $\gamma\epsilon\rho\rho\acute{\alpha}\omega$ I produce.

[2] $\Pi\epsilon\rho\acute{\iota}$ around, $\beta\lambda\tilde{\eta}\mu\alpha$ covering.

[3] $\Pi\lambda\eta\rho\tilde{\omega}\mu\alpha$ that which fills.

[4] $Ka\lambda\acute{\upsilon}\pi\tau\rho\alpha$ a cap.

7. *The system of receptacles for secretions.* Function: to receive the products of secretion of the plant.[1]

1. *The Epidermal System.*

While such plants and parts of plants which consist of a single
cell, or of but a single layer, do not possess an **epidermis**,[2] a
more or less distinct development of epidermal layers appears
already in the thallophytes and cormophytes, which have the
thickness of but a few layers of cells. The cells in these layers
become for the most part smaller and more thick-walled toward
the exterior, and often colored, even though the formation of a
true epidermis is not yet effected (*Secale cornutum, Fucus vesicu-
losus, Cetraria, Usnea, Sphærococcus,* the small stems of
mosses).

It is only in the higher plants that, even in the youngest
stages, a true epidermis is formed, and this is the first of any
system of tissue which is sharply defined, morphologically, from
all the others.

It consists in most cases of a row of tabular or plate-like cells,
laterally united without intervening spaces, which in the organs
of dicotyledons are, as a rule, of quadratic form, in the elongated leaves and stems of monocotyledons, however, are mostly
extended in the direction of the axis of the organ (distinct upon
surface sections). On many roots the epidermis is also sharply
defined from the other tissue by a different color and strongly
sinuous outer walls. Such an epidermis, which we meet with,
for example, on the rootlets of *Helleborus niger* and *Veratrum*

---

[1] Sachs classifies the tissues as follows: epidermal tissue, fascicular
tissue, and fundamental tissue. In the above division we adhere to the
classification of Haberlandt (" Physiologische Pflanzenanatomie," Leipzig, 1884), which is based upon Schwendener's principles. Nevertheless
the expression " fundamental tissue " (filling tissue) may often be permitted in the following pages on account of its brevity, notwithstanding
the fact that the fundamental tissue comprises the most varied forms
of tissue.

[2] 'Επί upon, and δέρμα skin.

album, was formerly termed *epiblema* (Figs. 84, 119, 120, 121, 122).

Occasionally, however, the epidermis consists of several layers, as, for example, in *Macis* (Fig. 85), and in many leaves (*Ficus*). This multiple epidermis, which is also termed *hypoderma,*[1] consists, in the case of delicate organs, mostly of uniform, thin-

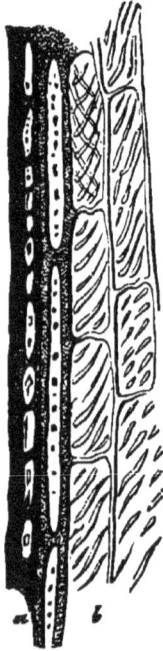
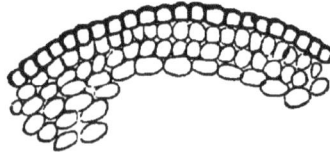

FIG. 83.   FIG. 84.

FIG. 83.—Longitudinal section through the outermost layer of *Vanilla; a,* epidermal cells, containing crystals of vanillin: *b,* cells with spiral fibres.

FIG. 84.—Epidermal cells of a root (epiblema) on a transverse section; the dark cells represent the epiblema.

walled or slightly thickened cells (leaves of the Piperaceæ, *Chavica* and *Peperomia,* of the Begoniaceæ, and of species of

[1] 'Ύπο under, and δέρμα skin. We use the word only for the true multiple epidermis, not for the layers (collenchyma, bast-fibres) which impart strength to the single-rowed epidermis.

12

Ficus); the outermost row, however, is also here, for the most part, somewhat differently formed.

In most fruits and seeds only the outermost row of cells be-

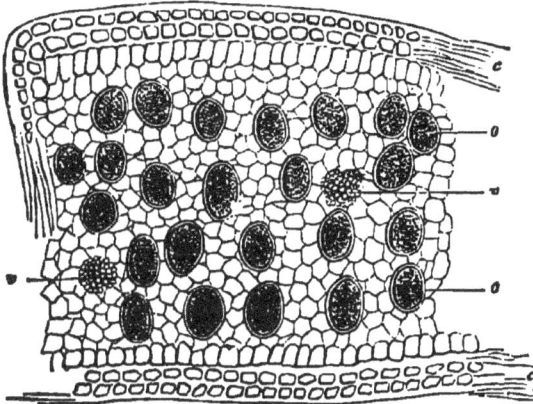

FIG. 85.—Transverse section through *Macis;* c, epidermis; o, oil-cells; v, fibro-vascular bundles.

longs to the epidermis, and in this case, to speak of a multiple epidermis, is incorrect.

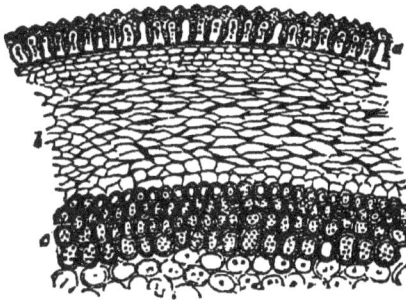

FIG. 86.—Rind of the fruit of *Colocynth* (in the commercial fruit usually removed by paring); a, epidermis; b, parenchyma; c, sclerenchymatous layer, which mostly forms the outer surface of the pared fruit.

In the epidermis consisting of a single layer, the cell-walls are mostly firmer than in the tissues lying beneath it, and more strongly thickened on the outer side than on the inner [1] (Figs. 83, 85, 63, 109, 129, 155). Occasionally the outer wall is even of quite remarkable thickness (*Caryophylli, Macis,* Fig. 160). An example where this is not the case is presented by the epi-

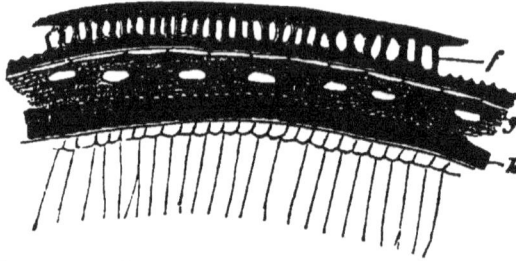

FIG. 87.—Transverse section through *Semen Paradisi* (Grains of Paradise); *f,* epidermis; *gh,* testa of the seed.

dermis of *Hyoscyamus* seeds. Here the outer wall of the epidermal cells of the testa is formed exclusively of the delicate cuticle.

The outer wall of the epidermis cells is always covered by the

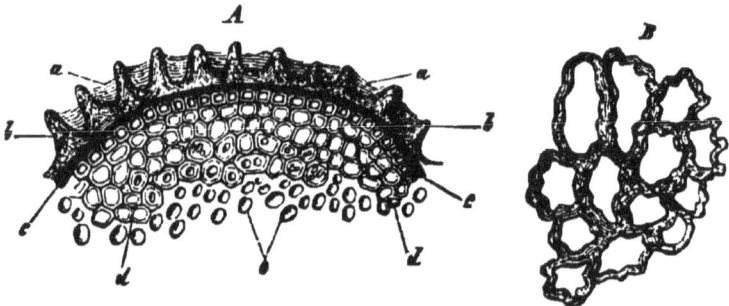

FIG. 88.—*Semen Hyoscyami.* *A.* Transverse section; *a,* cuticle; *b,* epidermis; *c,* testa; *d,* albumen cells; *o,* fatty oil. *B.* Tangential section through the cells of the epidermis.

cuticle (see page 161 and Figs. 155, 160, 63, 128, 161), usually a delicate [2] film, insoluble in sulphuric acid, impenetrable by water

[1] The delicate epidermal cells of the seeds of *Cydonia* and *Linum,* which produce mucilage, form an exception (Berg's " Atlas," xlvi., 122, 123).

[2] Compare Tschirch, " Ueber einige Beziehungen des anatomischen Baues der Assimilationsorgane zu Klima und Standort," Linnæa, ix. (1881), 139.

and aqueous vapor, and which for the previously described reasons is primarily adapted to fulfil the function of the epidermis namely, protection against too strong evaporation.

The cuticle, which is often directly visible upon transverse sections (Figs. 160, 161, 155), or is easily rendered visible by dilute chromic acid, sulphuric acid, iodine, or potassa, covers all the organs of the plant which are exposed to the air, and, in the case of favorable objects, can often be removed as a coherent film.

If the outer wall is thin, as for instance in the leaves of the indigenous foliage trees and all the officinal leaves, then the cuticle is in direct connection with the cellulose layer (Fig. 155); if, however, the outer wall is very strongly thickened, the intervening layers of the outer membrane of the epidermal cells are for the most part cuticularized (cuticular layers, Fig. 161 *c s*, 63), that is, they have become more similar to the cuticle

FIG. 89.—Transverse section through Fig. 88 A, more highly magnified.

itself by the deposition of cutin. These cuticular layers often project in a somewhat cone-like form towards the interior (Figs. 161, 63).

The outer wall is frequently somewhat arched in an outward direction (Fig. 155). These outward arches may become definitely shaped processes, whereby the exterior surface acquires a pitted appearance. The same appearance may, however, also be produced by the prominent development of thick lateral walls (testa of the seed of *Hyoscyamus*, Fig. 88, and the leaves of *Gentiana cruciata*). In the latter case, the outer wall is always very thin, and indented (Figs. 88 and 89). Occasionally the surface delineation is also produced by the projection of sharply circumscribed groups of epidermal cells.

The outer wall (like the thick lateral walls, Figs. 88, 89) shows, as a rule, a distinct stratification.

While the outer walls of the epidermal cells, as a rule, are thick, the lateral walls are mostly thin. The Figs. 63, 109, 129 and 155, therefore, reproduce the type of epidermal cells.

Upon surface sections, the lateral walls appear in many cases sinuous, as, for instance, in all leaves of the Gramineæ (Figs. 88 B, 90, 154, 157, 158), so that the individual epidermal cells which are provided with many protuberances fit into each other in a tooth-like manner (many corolla leaves, and the epidermis of *Semen Stramonii*, Fig. 90).

As a rule, the contents of the epidermal cells are colorless,

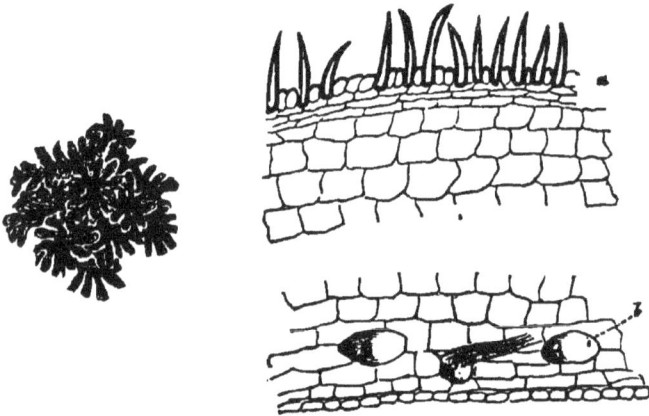

FIG. 90.   FIG. 91.

FIG. 90.—Tangential section through the epidermis of *Semen Stramonii*.
FIG. 91.—Transverse section through Chinese galls; a, epidermis, the cells of which frequently grow out in the form of simple hairs; b, lacticiferous cells.

without chlorophyll; occasionally, however, coloring matters appear in them, dissolved in the cell-sap. Colored epidermal cells produce, for example, the red color of many stems (*buckwheat, Ricinus*) and other organs (the *apple*). In the red potatoes, the coloring matter is contained in cells lying beneath the cork.

It has already been mentioned (page 180) that the epidermal cells often project outward. If these protuberances become larger, hair formations [1] or **trichomes** [2] are produced. These are found in the most simple, unicellular form, in *Stipites Dulcamaræ*, on *Herba Lobeliæ* and the *Chinese galls* (Fig. 91), and upon *cotton-seed* (Figs. 92, 114); the root-hairs (*Rad. Sarsaparillæ*, Fig. 126) also belong here.

The hairs are occasionally very long (flower buds of *Althæa rosea*).[3] Long, sharp, silicified hairs are termed *prickles* (stinging hairs of the nettle, Fig. 93 *a*). Firmer, short, non-secreting trichomes are called *bristles*. Of the latter kind are, among others, the hairs of *Nux vomica* (Fig. 93 *b*), of *anise* (Fig. 94).

The hair formations do not, however, always remain simply hair-shaped. Many of them assume other forms (that of a star, shield, or head, Fig. 95), throw out branches and become

FIG. 92.—Hairs of cotton.

multicellular.[4] Flatly expanded multicellular hairs (chaffy hairs), such as are of frequent occurrence in ferns (for instance, in *Aspidium Filix mas*) form the so-called Pengawar Djambi.

If the terminal cell of a multicellular trichome becomes ex-

[1] Compare Weiss, "Die Pflanzenhaare," in Nos. iv. and v. of the "Botanische Untersuchungen," of Karsten, 1867. Rauter, *loc. cit.*, 81. Martinet, Annal. d. sciences natur., xiv. (1872), 91–232. Paschkis, "Pharmacognostische Beiträge." Zeitschr. d. allg. œsterreich. Apothekervereins, 1880, Nos. xxvii. and xxviii. Hanstein, Bot. Zeit., 1868, 725. De Bary, "Anatomie," p. 61, where the literature is given to the year 1877.

[2] Θρίξ, τριχός hair.

[3] Sachs, "Lehrbuch der Botanik" (iv.), 101.

[4] The unicellular climbing hairs of the *hop*, reposing upon a multicellular cushion, are not of this kind.

panded in a head-like form, the formation of daughter cells frequently occurs, with the simultaneous secretion (for instance,

 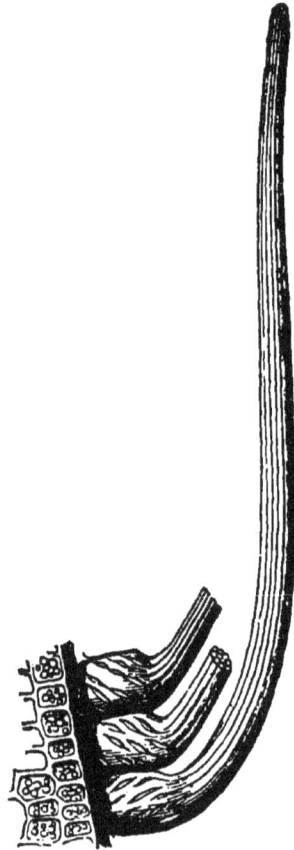

FIG. 98 *a*.   FIG. 98 *b*.

FIG. 98 *a*.—Stinging hair of the *nettle*, with a small head. The protoplasm of the hair circulating in currents (the direction of the currents indicated by arrows).

FIG. 98 *b*.—Hairs of the epidermis of *Nux vomica* (Berg).

in the Labiatæ) of a balsam or volatile oil, which is often accompanied by the formation of mucilage[1] (Figs. 96, 129,

[1] The morphology and the nature of the development of secreting

154). Such hairs are termed *glandular hairs* or colleters.[1] To this class belong also the glands of *Dictamnus*.[2]

In the species of *Cistus* of the Mediterranean flora, these hair formations secreting resin[3] are so numerous and so productive that, for example, the product of *Cistus ladaniferus* has

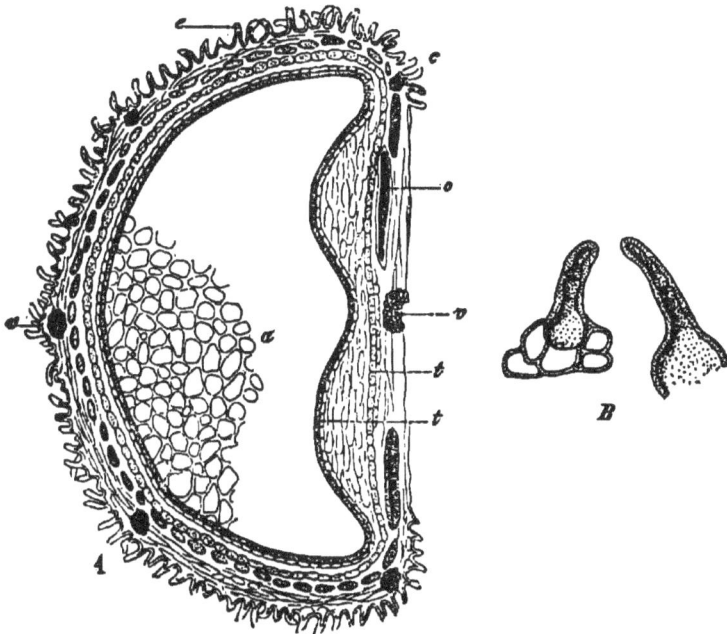

FIG. 94.—*A*, Transverse section through *Fructus Anisi. e*, epidermis, clothed with hairs; *cc*, commissural surface; *o*, oil spaces; *t*, coating of the fruit; *t'* (lower *t*), seed coat; *v*, fibro-vascular bundles (ribs, costæ); *a*, albumen of the seed, the parenchyma of which is indicated by but a few cells. *B*, Hairs, more highly magnified.

been collected in the islands of Candia and Cyprus from ancient times and employed for fumigating purposes. This *ladanum*

trichomes has been described by Hanstein (Bot. Zeit., 1868, 747); compare also De Bary, " Anatomie."

[1] Κολλητός glued together.

[2] Meyen, "Secretionsorgane der Pflanzen," Berlin, 1837; Plate i., Figs. 28 and 29. De Bary, " Anatomie," p. 78.

[3] De Bary, loc. cit., p. 99, Fig. 36.

resin is probably the only example of such a drug originating from trichomes.[1]

In *kamala* and *lupulin*, which, according to the nature of their development, should also be classed with the trichomes,[2] the formation of resin predominates; and in *kamala* oil is entirely wanting.

Glands which secrete wax are found on the leaves of *Globu-*

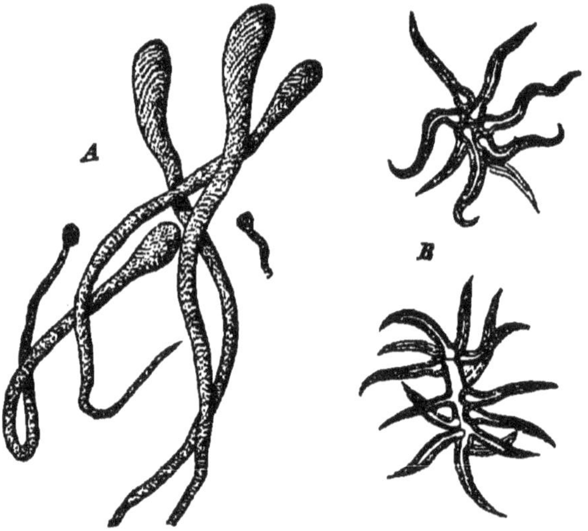

Fig. 95.—*Flores Verbasci.* *A*, Band-like, soft, club-shaped hairs of the three shorter stamens, covered with exceedingly fine, spirally-arranged, projecting points. *B*. Stellate hairs from the base of the folds of the corolla.

laria Alypum L. and other species,[3] those secreting nectar in *Melampyrum*.[4]

[1] Compare Thiselton Dyer, Pharm. Jour., xv. (1884), 301.

[2] With regard to the development of the lupulin glands, compare Holzner, "Entwickelung der Trichome der Hopfendolden." Bayer. Bierbrauer, 1877, No. 19. Rauter, "Denkschr. d. Wiener Akad.," 1870, p. 31. Luerssen, "Medizin.-Pharmaceutische Botanik," ii., p. 527. Harz, "Samenkunde," ii., 896.

[3] Heckel et Schlagdenhauffen, Comptes rendus, 95 (1882), 91.

[4] Rathay, "Ueber nectarabsondernde Trichome einiger Melampyrum-Arten." Wiener Akademie, 1880.

The form and size of the hairs, as also the relative thickness of the wall compared with the lumen or cavity, afford in some cases good points of discrimination for the recognition of adulterations in foods and alimentary substances. Thus Wittmack[1] distinguishes wheat and rye flour by the fragmentary hairs of the so-called coma, which always occur therein in small amount; and upon the form of the hairs Bell[2] bases the distinction of *tea, elder, willow* and *black-thorn leaves* (*Prunus spinosa* Lin.). The physiological function of the hairs of organs of assimilation (leaves), which when old mostly contain air, is to diminish the extent of transpiration. The hairs of seeds, which are often

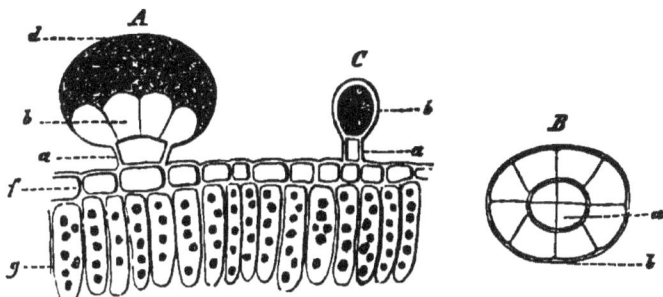

FIG. 96.—Oil glands of the Labiatæ, e. g., of *Rosmarinus.* *A,* Longitudinal section of a large gland. *a,* stem-cell; *b,* eight delicate-walled daughter cells which produce the volatile oil, by the escape of which the cuticle of the parent cell, *d,* becomes expanded; *f,* epidermis of the leaf upon which the gland is formed; *g,* palisade-cells; *e,* a small gland. *B.* Transverse section of Fig. *A.* Compare also De Bary, "Anatomie," Fig. 89.

feather-like, are means of distribution; the firm hairs of climbing plants are organs which serve to fasten them.

Some prickles, for example in *Rubus,* are, like the hairs, also of epidermal origin, and are thus trichomes. In the formation of most of the true prickles or *outgrowths,* however, the tissues beneath the epidermis, and even the vascular bundles, also participate (*Rosa, Smilax*[3]).

[1] "Anleitung zur Erkennung organischer und anorganischer Beimengungen im Roggen- und Weizenmehl." Leipzig, 1884.
[2] Bell, "Die Analyse der Nahrungsmittel," i. (Berlin, 1882), 36.
[3] De Bary, "Anatomie," p. 61, where the literature is given.

Finally, the so-called *inner hairs* may also be mentioned, which occasionally penetrate into the air-cavities (star-shaped hairs of the *Nymphæ*, glandular hairs of *Aspidium Filix mas*, Fig. 162).

The epidermis no longer suffices for older plant organs of several years' growth, since it is a much too delicate tissue (for instance, for the stems and branches), and, as permanent tissue, is not capable of keeping pace with the growth in thickness. In these organs, therefore, there is formed beneath the epidermis, and mostly independent thereof, another tissue, the **periderm.**[1] The latter consists of a permanent tissue, the **cork,** and a formative tissue, the **phellogen,**[2] or cork-cambium.

The phellogen, by a tangential division of its cells, forms the

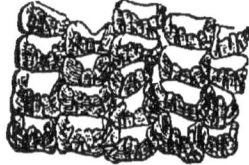

Fig. 97.—Cells from the phelloderm of the bark of *Canella alba* (J. Moeller).

cork-cells; but it is also capable of contributing to the increase of the bark parenchyma by the formation of parenchymatous elements. Sanio[3] terms the aggregate of cells which are thus produced, and occasionally thickened on one side (Figs. 97, 149) **phelloderm**[4] or cork layer of the bark.

[1] Περί around, and δέρμα skin.
[2] Φελλός cork, and γεννάω I produce.
[3] "Bau und Entwickelung des Korkes," Pringsheim's Jahrb., ii., (1860), 47. H. v. Mohl, "Entwickelung des Korkes und der Borke der baumartigen Dicotylen" (1836). "Verm. Schrift.," Tübingen, 1845, 225. F. von Höhnel, "Ueber Kork und verkorkte Gewebe überhaupt." Sitzungsberichte der Wiener Akademie, 76 (1877). Hanstein, "Bau und Entwickelung der Baumrinde." Berlin, 1853. Hofmeister, "Handbuch der phys. Botanik," i., 252. De Bary, "Anatomie."
[4] Φελλός cork, and δέρμα skin.

The location of the phellogen is not exclusively confined to the region directly under or within the epidermis, but may also develop itself in the form of bands and stripes in the fundamental tissue of many barks, or even in the phloëm layer. Outside of such layers of *internal cork* (Fig. 98), the most various tissues of the bark may therefore be represented, according to the depth at which they lie (phloëm elements, Fig. 100, bastcells, Fig. 98, stone-cells, Fig. 101, and, indeed, even resin-

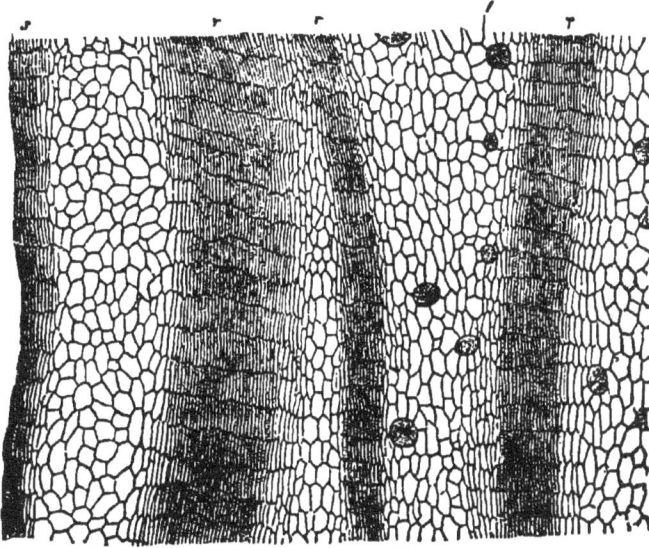

FIG. 98.—Transverse section through the bork of *Cinchona Calisaya. s*, outermost cork layer; *r*, cork-bands in the inner tissue; *l*, bast-cells (Berg.).

canals and oil-spaces, Figs. 99, 100); they become pushed out of the course of circulation of the sap by these lamellæ of internal cork, and are either thrown off as scales (in a very handsome manner in the *Platanus* and *Eucalyptus*), or they still remain for a long time united with the stem, and appear severed and torn only in consequence of the growth in thickness (*bork, rhytidoma* [1]).

[1] 'Pυrίs, ῥυτίδος fold, wrinkle, and δωμάω I build.

The bork of our foliage trees, for example of the *oak*, contains, in addition to cork-cells, all the elements of the outer bark (parenchyma, stone-cells).

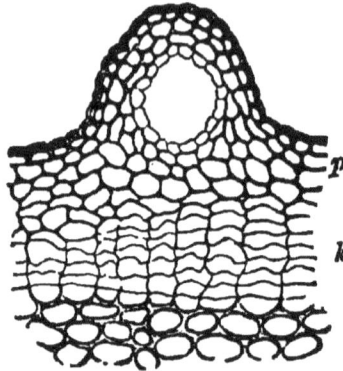

FIG. 99.—Transverse section through an older internode of *Juniperus communis* L.; *p*, outer layer of the primary bark, with a resin canal; *k*, interior cork (J. Moeller).

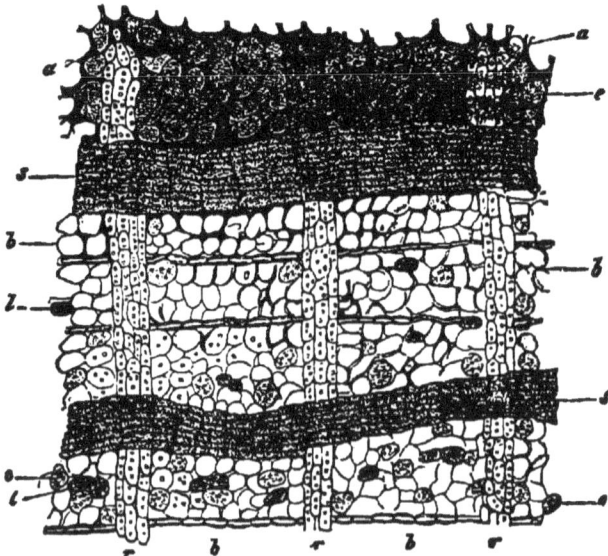

FIG. 100.—Bork of *Cortex Sassafras radicis;* *aa*, decayed surface; *ss*, cork-bands ; *bb*, phloëm; *o*, oil cells; *r*, medullary rays.

Whether barks experience the formation of bork, or are provided with a simple covering of cork, appears to depend upon

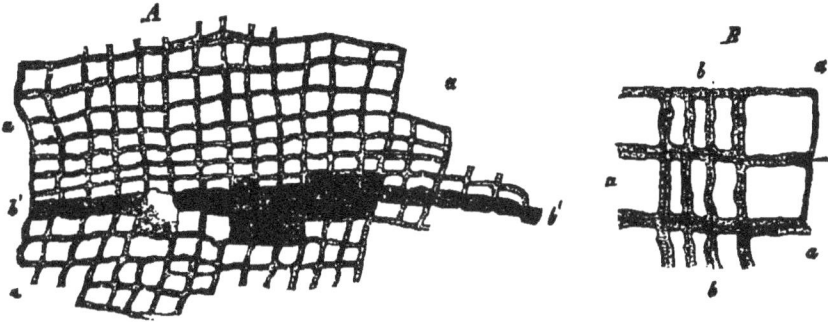

Fig. 101.—*A*, Cork of the *cork-oak; aa*, cork-cells; *bb*, stone-cells; *B*, more highly
magnified cork-cells.

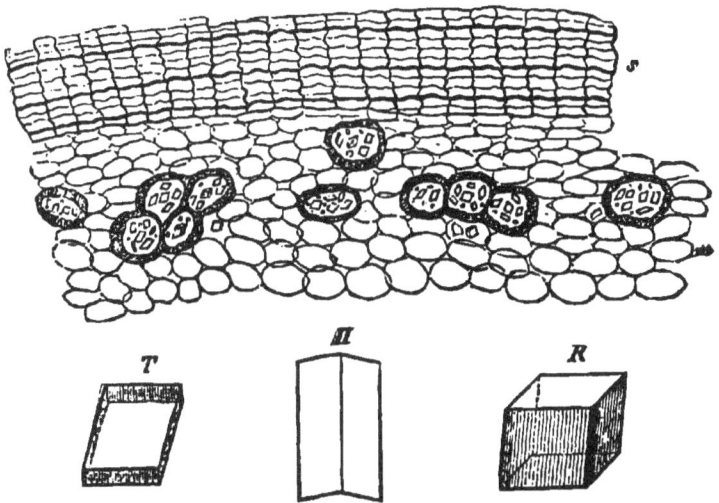

Fig. 102.—Cortical layer of *Radix Calumbœ; s*, cork; *m*, fundamental tissue, with dispersed stone-cells which inclose crystals of oxalate; *R*, *T*, More highly magnified
crystals; *H*, twin crystal.

the peculiarities of the species. In the *Cinchonas*, for example, bork is sometimes met with, and sometimes not. Roots are

also capable of forming bork, which may be very plainly observed, for instance, in *Radix Sassafras*.

If the periderm layers which are formed in the interior of the bark occupy only a part of the circumference, *scale-bork* is produced (*Robinia, Platanus, Cinchona, Pinus silvestris, Quercus*); if, however, the secondary periderm layers form parallel,

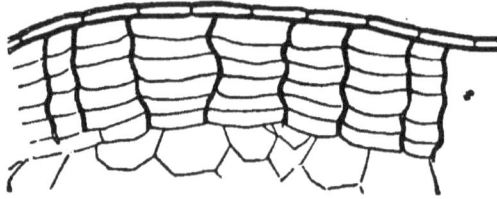

FIG. 103.—Transverse section of *Rhizoma Curcumæ*; *s*, cork.

closed rings, which embrace the entire circumference, hollow-cylindrical sections of bark are converted into bork, and *ringed-bork* is produced (*Vitis, Clematis*).

FIG. 104.—*Cortex Cascarillæ.* Cork layer and primary bark, with crystals of calcium oxalate and coloring matter.

Since the cork-cells are produced by a tangential division of mostly tangentially extended cells, they are flatly tabular and parallelopipedal. They are in unbroken connection with each other and contain air, never solid matter. As the divisions in the phellogen take place very regularly, the transverse walls of the cork-cells frequently traverse the entire cork tissue in one and the same line (Figs. 102, 103, 108).

The ordinary cork, of *Quercus Suber*, corresponds in its form to the above type.[1] Deviations from this fundamental form

Fɪɢ. 105.—*Cortex Guaiaci.* a, thickened cork-cells; a', phellogen layer; b, primary bark; c, sclerenchymatous layer.

depend upon a more undulating—though on a transverse section, generally radial—course of the transverse walls, or upon a

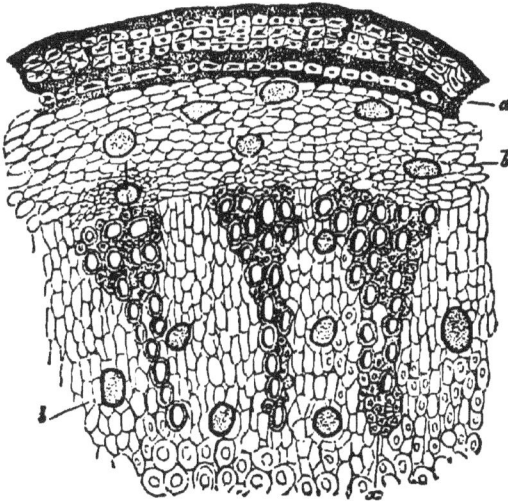

Fɪɢ. 106.—*Radix Pyrethri romani.* a, thickened cork-cells; b, oil-spaces; x, xylem rays (wood-bundles).

thickening of one or all sides (as seen in Figs. 102, 103, 108) of the ordinarily thin walls (Figs. 97, 104, 105).

[1] Yet in cork (bottle-cork), numerous stone-cells occur (Fig. 101).

The outermost layer is often in process of decay, as for instance in the *potato* (Fig. 108). While this perishes, new cork

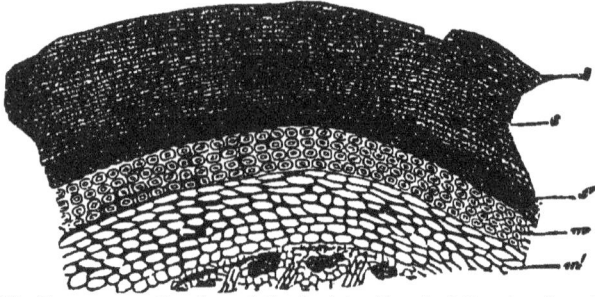

FIG. 107.—Transverse section through the bark (periderm) of Jamaica Quassia-wood
s, cork; *s'*, cork-cambium or phellogen; *m*, layer of the primary bark; *m'*, containing
crystals.

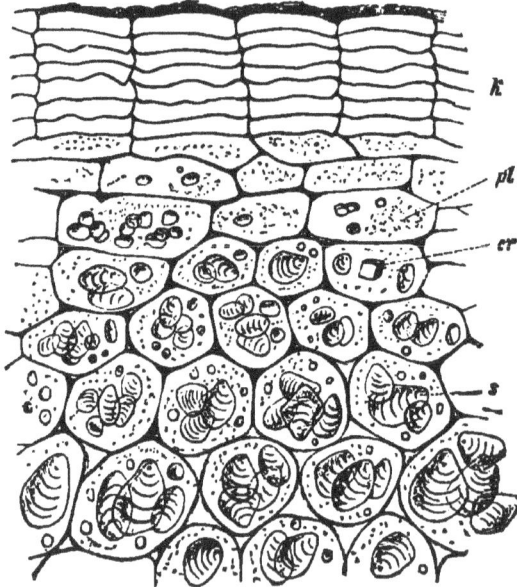

FIG. 108.—Transverse section through the outermost layer of a potato. *s*, starch
granules; *cr*, protein crystalloid; *pl*, protoplasm (Tschirch).

is continually formed from the interior (*Cortex Quassiæ jamai-
censis*, Fig. 107).

13

Even on leaves a local formation of cork occasionally occurs[1] (*Eucalyptus*, Fig. 128, *k*). The physiological function of cork is to protect the tissues lying beneath it from too great evaporation and from mechanical injury. The former function is presented in a very striking manner by the potato, in which the cork layers, from five to ten in number, cause the succulence of the entire inner tissue to be retained for a long time undiminished.

On wounds, the so-called *wound cork* is frequently formed (Fig. 109, *a*). This produces, in the same manner as the *thyllæ*[2] and the gum which incloses the vessels, a separation of the inner, uninjured tissues from the wounds.[3]

2. The Mechanical System.

Leaving out of consideration the lower plants, we find that all the higher plants which concern us are provided with peculiarly formed and characteristically arranged cells, the exclusive function of which, as Schwendener[4] has shown, is to impart to the plant the necessary solidity. While in young and still growing organs the **collenchyma**[5] (Fig. 109 *b*, 129 *coll*) represents the mechanical system, in older and matured organs this function is assumed by the **bast-cells**, or stereïds[6] (Figs. 110, 111).

The bast-cells, or the specifically mechanical elements of the matured plant, form very elongated cells, which are pointed at

[1] Bachmann, "Korkwucherungen auf den Blättern." Pringsheim's Jahrb., xii., 1880.

[2] Sachs, "Lehrbuch der Botanik," iv. (1874), pp. 27, 782. Weiss, "Anatomie," 21.

[3] With regard to the lenticels, see the chapter on the Aërating System.

[4] "Das mechanische Prinzip im anatomischen Baue der Monocotylen." Leipzig, 1874.

[5] Ambronn, "Ueber die Entwickelungsgeschichte und die mechanischen Eigenschaften des Collenchyms." Pringsheim's Jahrb., xii. E. Giltay, "Het Collenchym," Inaugural dissertation, Leyden, 1882.

[6] The tissue of bast-cells may be termed *stereom*. (The pleurenchyma of Meyen.)

both ends, provided with oblique pores (Fig. 110), often thick-

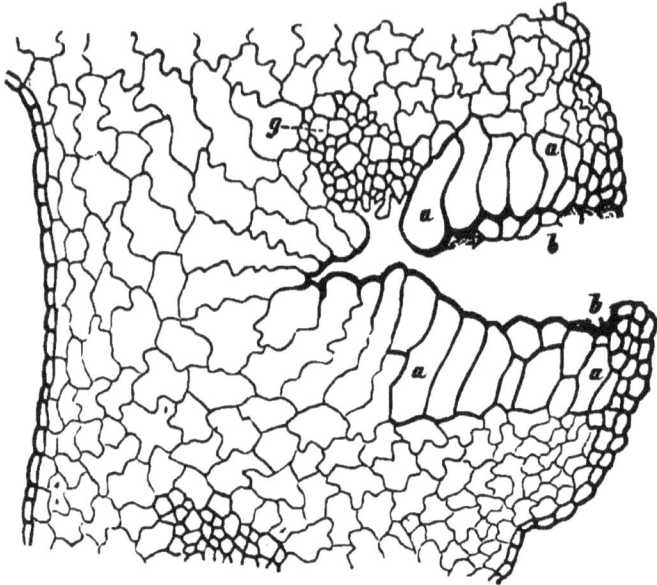

Fig. 109 *a*.—Transverse section through a fruit of the *Vanilla*, which, before separation from the plant, had become injured at the point *b–b* by the puncture of an insect. and the wound closed by wound-cork, *a–a; g*, vascular bundles (Tschirch). Compare *Pharm. Zeit.*, 1884, No. 22.

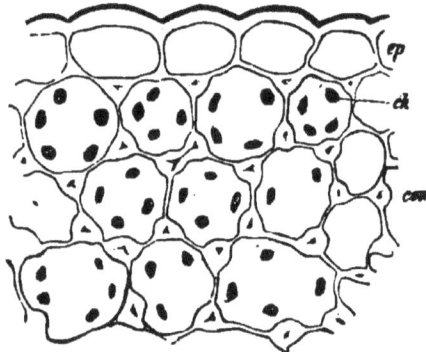

Fig. 109 *b*.—Collenchyma. *ep*, epidermis; *con*, collenchyma cells; *ch*. chlorophyll granules (Tschirch).

ened to the extent of the disappearance of the lumen or cavity, and contain air. They possess a supporting capacity which is almost equal to that of wrought-iron, and a tenfold greater ductility than the latter, and are therefore in themselves admirably adapted for the mechanical purposes of the plant.

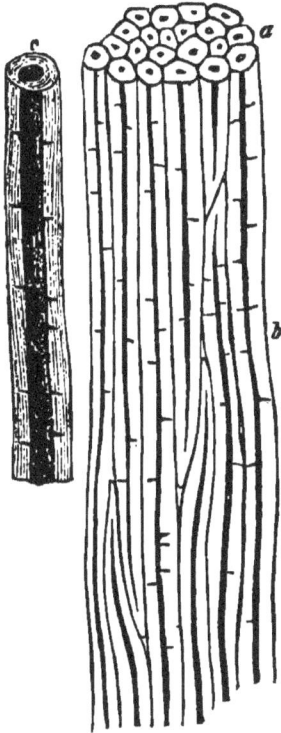 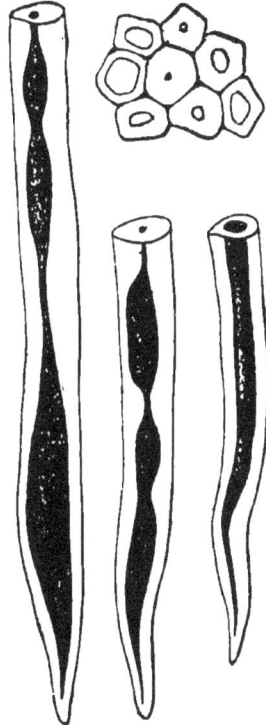

FIG. 110.   FIG. 111.

FIG. 110.—Typical bundle of bast-cells, at *a* in transverse section, at *b* in longitudinal section: *c*, section of a bast-cell, showing the striping of the membrane and oblique pits (Tschirch).

FIG. 111.—Bast-cells of *Corchorus olitorius* (*Jute*), with a lumen or central cavity of varying width, at the top in transverse, and below in longitudinal section (Tschirch).

Moreover, they are always united to form structures which are in nowise inferior to the best constructions of our engineers.[1]

[1] With regard to the firmness of bast-cells, compare Schwendener's

In the monocotyledons the bundles of bast-cells surround the
vascular bundles, either in a sickle-like manner (as in the stem
of the *maize*, Fig. 133), or lie embedded in the remaining
tissue, sometimes in the inner portion and sometimes in the
outer, as isolated bands, rings or bundles, according to their
function, or they surround the outer side as a connected coat[1]

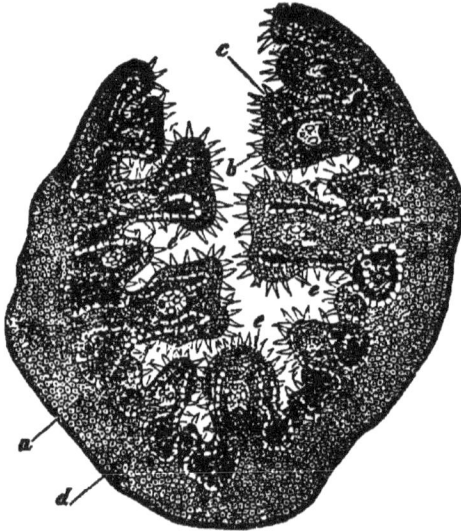

Fɪɢ 112.—Transverse section through an involute leaf of the *Alfa grass* or *Esparto*
(*Macrochloa tenacissima*). *a*, bast-cell coating of the outer (under) side; *d*, assimilating
tissue; *c*, prisms of the upper side; *b*, hairs (Tschirch). Compare *Pharm. Zeit.*, 1882,
No. 68.

principal work; furthermore, Th. v. Weinzierl, "Beiträge zur Lehre
von der Festigkeit und Elasticität vegetabilischer Gewebe." Sitzungsberichte der k. Akademie der Wissensch., Vienna, 1877, Vol. 76. F.
Lucas, "Beiträge zur Kenntniss der absoluten Festigkeit von Pflanzen-
geweben." Sitzungsberichte der Wiener Akademie, 1882, Vol. 85, and
1883, Vol. 87.

[1] In some of these cases (especially in the prairie grasses), they serve
in the mechanism which causes the involution of leaves (Tschirch,
Pringsh. Jahrb., xii.). In the dehiscence of fruits, and the phenomena
of torsion of many awns, mechanical cells are also concerned [Stein-
brinck, Inaugural dissertation, 1873, and "Berichte der Deutsch. Bot.

(*Alfa grass,* Fig. 112).   In the dicotyledons they are located in the bark (Fig. 113).

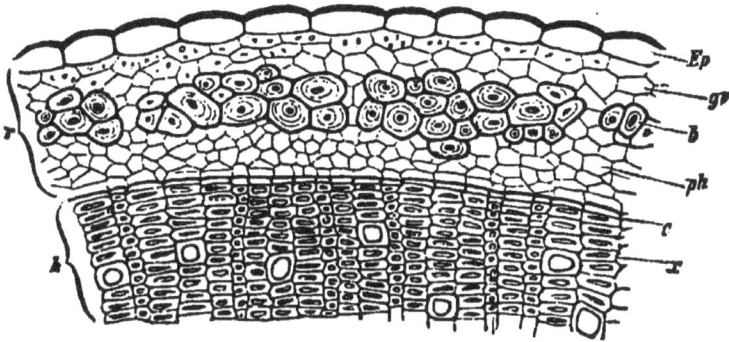

FIG. 113.—Transverse section through a stem of *Linum usitatissimum ; ep*, epidermis; *gr*, green bark; *b*, bast-cells (employed as flax); *ph*, phloëm (sieve portion); *c*, cambium; *x*, xylem (vascular portion); *r*, bark; *h*, wood (Tschirch).

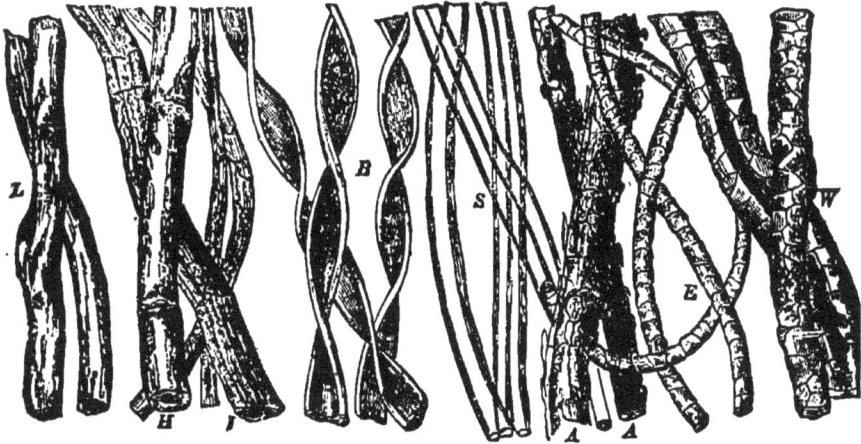

FIG. 114.—Illustrations of the more important fibres which are used technically. L. *flax fibres;* H. *hemp fibres;* J, *jute;* B, *cotton;* S, *silk;* A, *alpaca wool;* E, *Electoral wool;* W, *sheep's wool.*

Ges.," i. (1883), p. 27, and others.   Zimmermann, Pringsheim's Jahrbücher, xii. (1881), No. 4.]

The great mechanical service which the bast-cells are capable of was also recognized at an early period. The application of the bast-fibres of *hemp* and of *flax* (Fig. 113) for fabrics is a very ancient one.

The textile fibres which are practically employed may be grouped in the following manner (compare Fig. 114):

1. Animal fibres: (*a*) hairs, *Wool* (W); (*b*) threads, *Silk* (S).
2. Vegetable fibres:
    (*a*) hairs, *Cotton* (B).
    (*b*) bast-cells, *Flax* (L); *Hemp* (H); *Jute* (J); *Esparto; Manilla-hemp.*[1]

In addition to the bast-fibres, the **libriform cells** of the wood also assume, especially in the older stems of dicotyledons,

FIG. 115.—*A*, Sclerenchyma from the inner layer of the seed vessel of *Fructus Cocculi* (*Cocculus Indicus*). B, Some branched cells of the same, more highly magnified.

a mechanical function. The bast-cells which occur in the bark of dicotyledons (*hemp, flax,* Fig. 113, the *linden*) are only of service in imparting strength of flexure to the stem as long as the wood itself has not yet acquired sufficient strength.

In accordance with the various mechanical demands made of the plant, the structures in which the mechanical elements are

[1] Compare especially Wiesner, "Die Rohstoffe des Pflanzenreiches." Leipzig, 1873. Reissek, "Die Fasergewebe des Leines, Hanfes, der Nessel und Baumwolle." Denkschr. d. Wiener Akad., 1852. Berthold, "Ueber die mikroskopischen Merkmale der wichtigsten Pflanzenfasern." Zeitschr. f. Waarenkunde, 1883, No. 3; 4. Dorkoupil, "Materialien zu einem Lehrbuch der chemischen Technologie für Gewerbeschulen." Jahresber. d. Gewerbeschule Bistritz, 1882.

united are also very manifold. We may thus distinguish structures intended to protect the organs of the plant from the influence of bending, pressure, traction, or laceration at the
edges or margins (for instance, of leaves).

The stone-cells (sclereïds,[1] compare also pages 156 and 171),
as one of us (T.) has shown, also possess various mechanical
functions. In seeds which do not possess a thick-walled endosperm, there is found, for example, a hard endocarp, consisting of stone-cells (Figs. 115, 87). This often consists of very

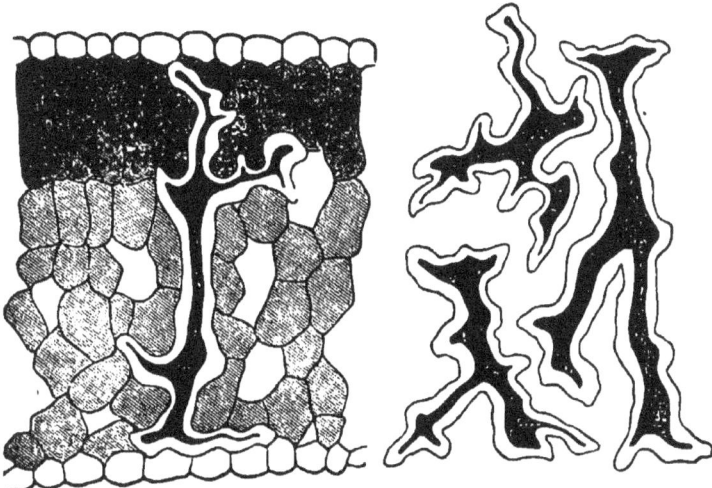

FIG. 116.—Transverse section through a *tea-leaf*, with the characteristic. branched
sclerenchyma cells. The palisade-parenchyma, which is richer in chlorophyll, is of
a darker color. At the right of the figure are sclereïds, isolated by maceration
(Tschirch).

variously adjusted rows of cells, and can therefore also endure
strong pressure and great expansion or straining (as by germination) without becoming ruptured. The function of the
sclereïds, which are so characteristic of *tea-leaves* (Fig. 116 [2]), is

[1] The tissue of stone-cells, in accordance with Mettenius, may be
termed *sclerenchyma* (see pages 156 and 172). Compare Tschirch,
" Beiträge zur Kenntniss des mechan. Gewebesystems." Pringsheim's
Jahrb., 1885.

[2] In young tea-leaves, however, these astrosclereïds are wanting.

doubtful. Occasionally the stone-cells contain crystals (Fig. 102) or other substances.

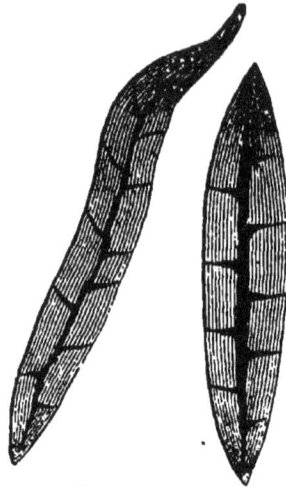

Fig. 117.—Short bast-cells from *Cinchona barks*.

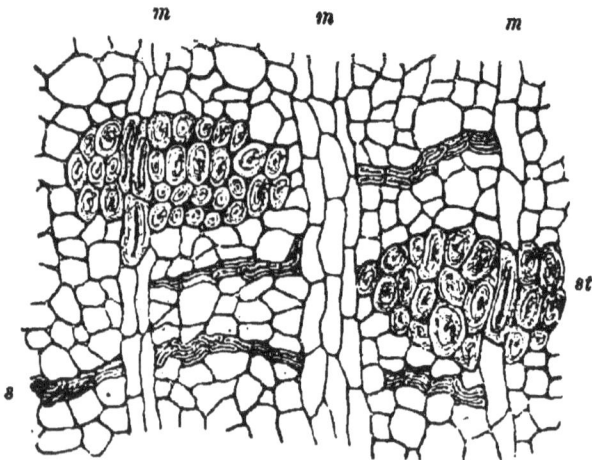

Fig. 118.—Transverse section through the sieve portion of *Cortex Coto*. *st*, staff-shaped bast-cells; *m*, partially sclerotized medullary rays; *s*, bundles of collapsed sieve-tubes (Moeller).

The function of the isolated bast-cells and stone-cells, for example in the *Cinchona barks* (Figs. 117, 145), in the *Pomegranate-root* (Berg's "Atlas," plate XL.), *Aconite root* (Fig. 121), *Simaruba, Coto bark* (Fig. 118), *Cort. Guaiaci* (Fig. 105), *Stipites Dulcamaræ* (Fig. 144), and *Oak-bark* is also still in doubt. Such cells, however, by the form of their transverse sections and their arrangement, present very useful points of discrimination in the characteristics of many drugs (*Cinchona* [1]).

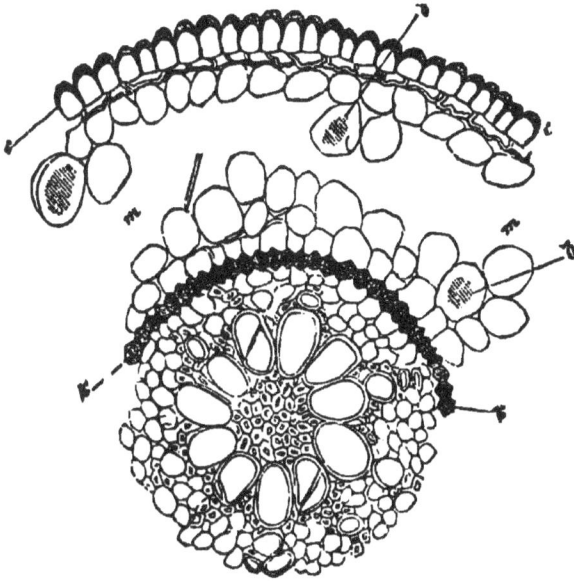

Fig. 119.—Transverse section of a rootlet of *Rhizoma Veratri*. *cc*, epiblema (p. 177); *m*, fundamental tissue, only partly represented in the drawing; *δ*, needle-shaped crystals; *kk*, nucleus-sheath.

The **nucleus sheath**, vascular-bundle sheath or protective sheath (endodermis [2]), which occurs in the root formations of

[1] Compare also in this connection, Koch, "Beiträge zur Anatomie der Gattung Cinchona," 8vo, pp. 35, 2 plates. Freiburg dissertation, 1884, especially page 20.

[2] Ἔνδον within, and δέρμα skin. Compare Schendener, "Die Schutzscheiden und ihre Verstärkungen," Abhandl. der Berliner Akademie, 1882.

vascular cryptogams, of monocotyledons and some dicotyledons,[1]

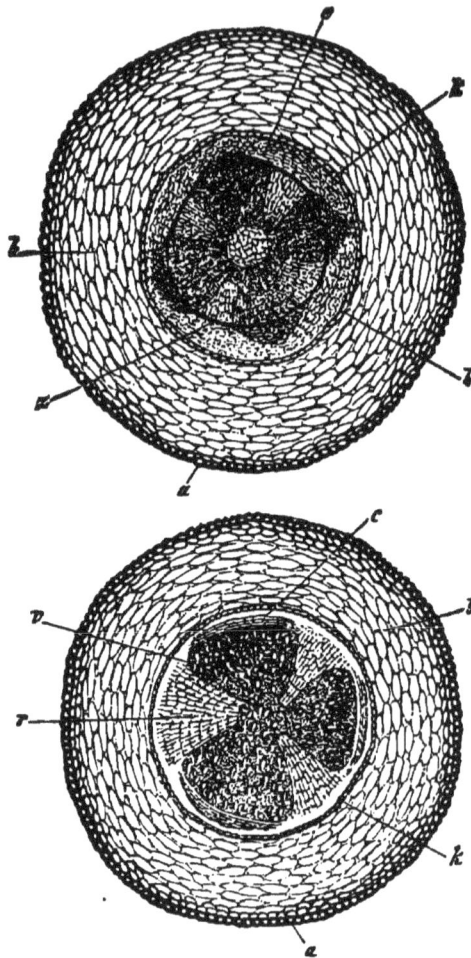

FIG. 120.—Transverse sections through rootlets of *Actæa spicata*. *a*, epiblema; *b*, fundamental tissue; *c*, phloëm; *e*, inner bark; *k*, nucleus-sheath; *z*, cambium; *r*, medullary rays; *v*, xylem.

[1] Compare C. van Wisselingh, "De Kernscheede bij de wortels der Phanerogamen," Amsterdam, J. Müller, 1884. (From: "Verslagen

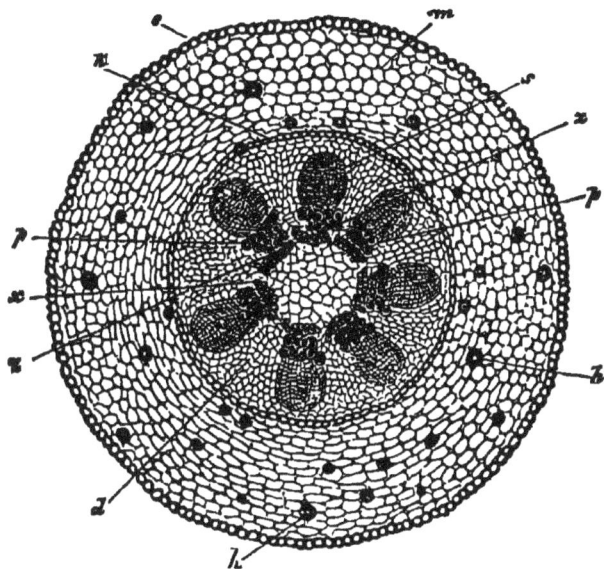

Fig. 121 B.

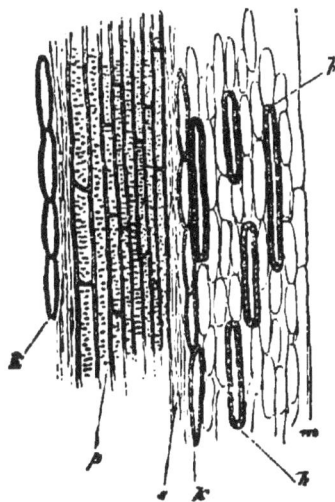

Fig. 121 C.

Fig. 121.—Rootlets of *Aconitum Napellus; B*, Transverse section; *C*, Longitudinal section in a radial direction; *e*, epiblema; *m*, fundamental tissue; *h*, stone-cells; *k*, endodermis; *p*, vascular bundles; *z*, xylem; *s*, phloëm.

en Mededeelingen der k. Academie van Wetenschappen," Afdeeling Natuurkunde, 3de Reeks, Deel i., pages 141 to 178, with 1 plate).

and likewise possesses mechanical functions, is also to be considered here. In these organs, namely, or at least in those cases which more nearly concern us, all the bundles, or a predominating number of them, are inclosed by a single row of cells (*Sarsaparilla*), or, by a layer, the endodermis, which is seen to

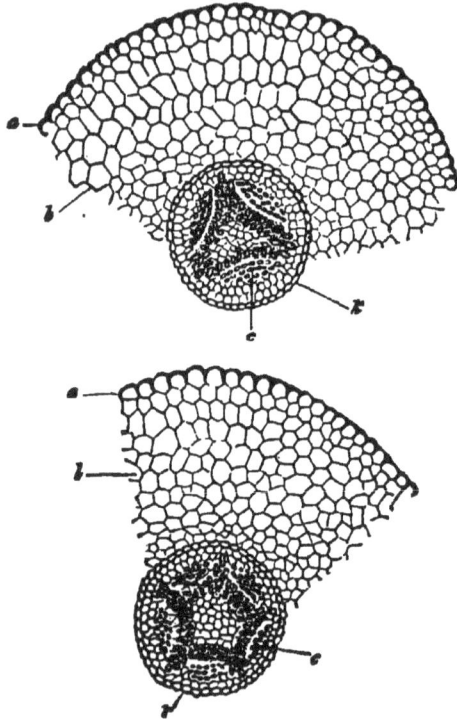

Fig. 122.—Rootlet from *Helleborus viridis;* a, epiblema; b, fundamental tissue: c, central bundle; k, endodermis.

be narrow on a transverse section and consists of but a few cells (*Galanga*). All the vascular bundles are located within the endodermis, for instance, in *Radix Sarsaparillæ*, *Rhizoma Caricis*, in the rootlets of *Actæa spicata* (Fig. 120), *Aconitum* (Fig. 121), *Helleborus* (Fig. 122), *Serpentaria, Valeriana*, and *Veratrum* (Fig. 119). On the other hand, in *Rhizoma Calami, Rhiz.*

Graminis, Rhiz. Iridis, Rhiz. Curcumæ, Galangæ, Zedoariæ
and *Zingiberis* the fundamental tissue outside of the sheath also
contains isolated bundles.

The endodermis, for example in *Sarsaparilla*, is composed
of prismatic cells, greatly elongated in the direction of the axis

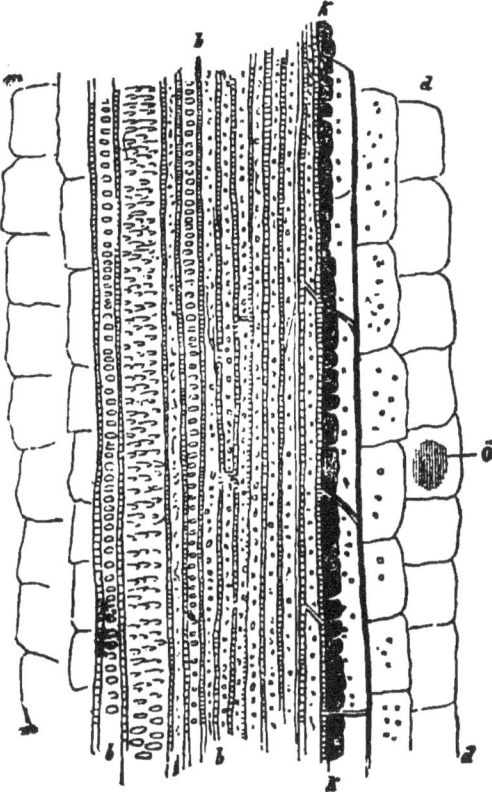

Fig. 123.—Radial longitudinal section through the endodermis (k) of sarsaparilla; d, middle bark; b, bundles; m, medulla; o, calcium oxalate.

(Fig. 123), forming a tube or sheath, located centrally in the
fundamental tissue, and containing within it the bundles. In
some varieties of *Sarsaparilla*, as also in *Rhizoma Graminis*, the
bundles are crowded against the sheath in the form of a compact

circle, in other cases they are dispersed, as is the case in *Rhizoma Veratri* and *Rhizoma Caricis*, or to a still greater extent in *Tuber Aconiti*, or but a single central bundle is present, as in the rootlets of *Veratrum*.

The cells of the sheath are not always elongated, but often

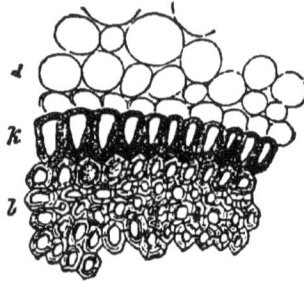

FIG. 124.—Transverse section through the endodermis of *Vera-Cruz Sarsaparilla*.

nearly cubical or only slightly extended. They are also often thin-walled, contain starch, and are then called *starch-layer* or *starch-sheath.*

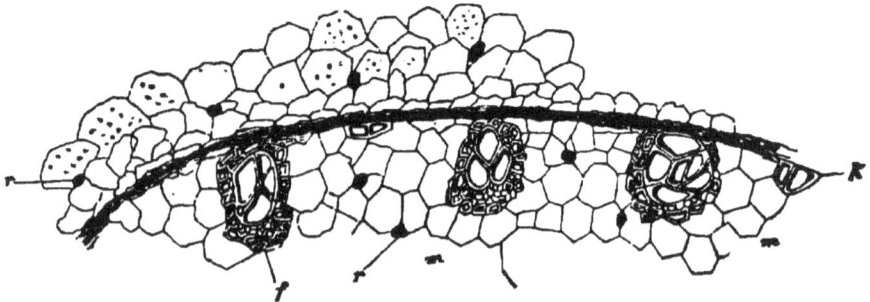

FIG. 125.—Transverse section through the endodermis (*k*) of *Rhizoma Galangæ; f*, fibro-vascular bundles; *r*, resin-cells; *m*, fundamental tissue.

The walls of the endodermis which are directed toward the axis are usually thickened; in the lateral walls this is also sometimes the case, so that the lumen or cavity, for instance in the Vera Cruz *Sarsaparilla* (Fig. 124), becomes very much contracted. The transverse sections of these cells of the nucleus-sheath therefore appear differently, according to the thickness of the thickened layers, and thereby afford serviceable characteristics for the recognition of the several varieties of a drug.[1] While the nucleus-sheaths in most of the examples that have been cited are built up of a single row of uniform cells, the root-stocks of the Zingiberaceæ deviate considerably in this respect. Indeed, the endodermis of the rhizomes of *Curcuma*, *Galanga*, *Zedoaria*, and *Zingiber* is composed of several rows of cells [2] (Fig. 125).

3. *The Absorbing System.*

The absorption of inorganic salts from the soil is effected by the aid of the roots and especially by means of the *root-hairs*. The latter, which are true trichomes, by forming manifold protuberances, become most intimately attached by their growth to the particles of the soil.

Root-hairs are found on but few officinal roots (for instance, *Sarsaparilla*, Fig. 126). In most cases they are broken off in the process of unearthing them, or they may have been already absent at the time of collection, since the formation of root-hairs only takes place in definite and young parts of the root.

For the absorption of organic nourishment, the phanerogamous parasites penetrate the host-plant by means of the so-called *haustoria* (as in the case of *Cuscuta*). To these haustoria correspond the surfaces of the cells lying close to the endosperm, which consist mostly of palisade-shaped cells with protuberances resembling root-hairs, and which are especially met with on the scutellum of the Gramineæ. They serve for imbibing the reserve substances.

In order to convey nourishment to the embryo during germina-

---

[1] Compare Schleiden, "Beiträge zur Kenntniss der Sarsaparilla." Archiv der Pharm., 1847. Arthur Meyer, *Ibid.*, 218 (1881), p. 280 *et seq.* Berg's "Atlas," Plate iv. Flückiger, "Pharmakognosie," p. 295.

[2] Compare Arthur Meyer, Archiv der Pharm., 218 (1881), p. 419.

tion, a longitudinal cleft is also occasionally found in 'the endosperm (*Strychnos Nux vomica, Coffea* ¹).

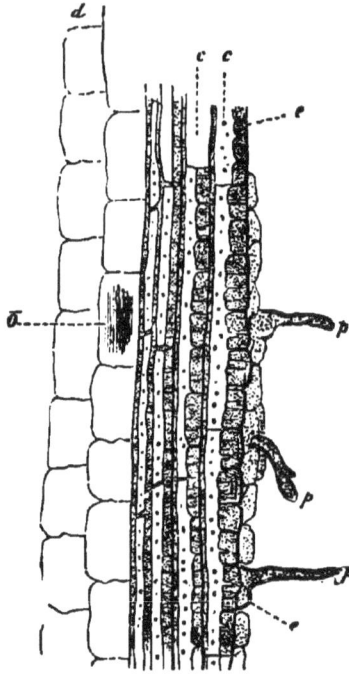

FIG. 126.—Longitudinal section through *Radix Sarsaparillæ; e*, epiblema; *p*, hairs; *cc'*, cells of the bark, which are thickened on one side; *d*, parenchyma; *o*, calcium oxalate.

4. The Assimilating System.

The assimilating tissue serves primarily for the formation of organic substance from carbonic acid and water under the influence of light, to which procedure the name of **assimilation** has been given. This tissue is filled with chlorophyll granules (compare page 100), and its cells possess forms which tend, as far as possible, to the transmission of light on all sides, and the rapid removal of the products of assimilation.²

¹ Jäger, " Endosperm der Coffea." Bot. Zeit., 1881, p. 336.
² Compare in this connection Haberlandt, in Pringsheim's Jahrb., xiii. (1881).

14

The surface of the leaf which is chiefly exposed to the light becomes the assimilating side. With *bifacial*[1] leaves, that is, such as are flatly expanded, and the upper and under surface of which is differently developed (as in the heart-shaped *Fol. Eucalypti*, Figs. 2 *b*, 127, 129, *Lactuca Sativa*), this is the upper surface; with *centric* leaves, that is, such as are placed vertically, and both sides of which are equally constructed (*Fol. Eucalypti*, sabre-shaped, Figs. 2 *a*, 128, *Lactuca Scariola*), it is both sides.

The cells of the assimilating side, which is always of a darker green color, are replete with numerous chlorophyll granules, located along the walls, and are extended in a palisade-like man-

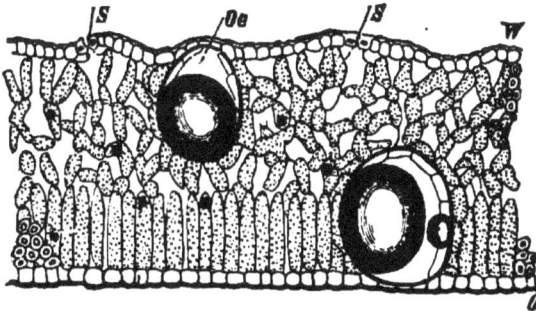

FIG. 127.—Transverse section through a heart-shaped (bifacial) leaf of *Eucalyptus globulus*; *oe*, oil-space; *s*. stomata; *w*, under surface; *o*, upper surface. After Tschirch, *Pharm. Zeit.*, 1881, No. 88.

ner more or less perpendicular to the vertical axis of the leaf (Figs. 96 *g*, 127, 128, 129 *pal*, palisade[2] parenchyma).

The special development of palisade parenchyma remains unaffected only in typical shade-plants.[3] On the other hand, in all centrically constructed leaves, both sides are provided with

[1] *Bis*, twofold, and *facies*, side.

[2] From the French word *palissade*, and this from the Latin masculine palus (not pallus!), therefore not pallisade, as it is often incorrectly written. (This applies more especially to the German orthography. F. B. P.)

[3] *Globularia Alypum* L. and other species present a notable example of an homogeneous leaf tissue without a palisade layer.

palisade cells (Fig. 128). The assimilating surface is often increased by a falling off of the membrane (as in the needle-shaped
leaves of the Coniferæ). In order to be able to take up the
products which are formed and to conduct them rapidly, the
palisade cells are now and then located upon funnel-shaped col-
lecting cells, which are in connection with the proper cellular threads (vascular bundles); the latter running into the
nerves of the leaves as a much branched radiating system, with
extremely fine terminations (Fig. 130).

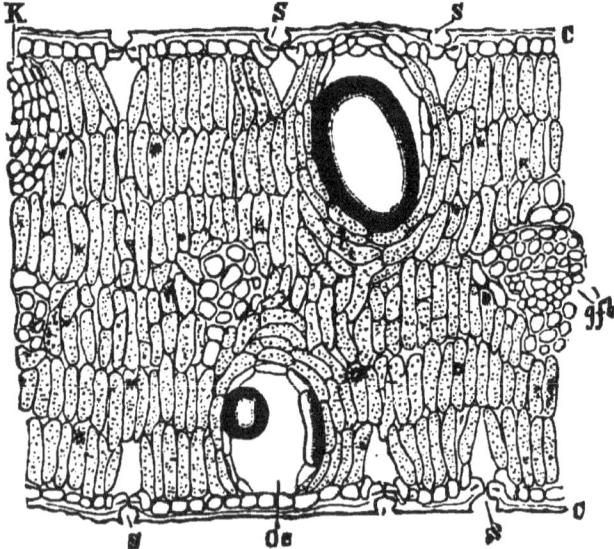

FIG. 128.—Transverse section through a sabre-shaped (centric) leaf of *Eucalyptus
globulus;* *oe,* oil-spaces, with drops of oil; *gfb,* vascular bundle; *s,* stomata; *k.* corky
growths; *c,* cuticle (Tschirch). Compare also Fig. 2 *a* and *b.*

The under side of the leaf, which is always of a lighter green
color, contains much less chlorophyll than the upper side, and
is traversed by wide air-canals (spongy parenchyma, leaf-
merenchyma, Fig. 129 *sch*).

The entire interior of the leaf, with the exception of the
vascular bundles, which is inclosed by the two epidermal sides,
is termed the *mesophyll.*[1]

[1] Μέσος in the middle, φύλλον leaf.

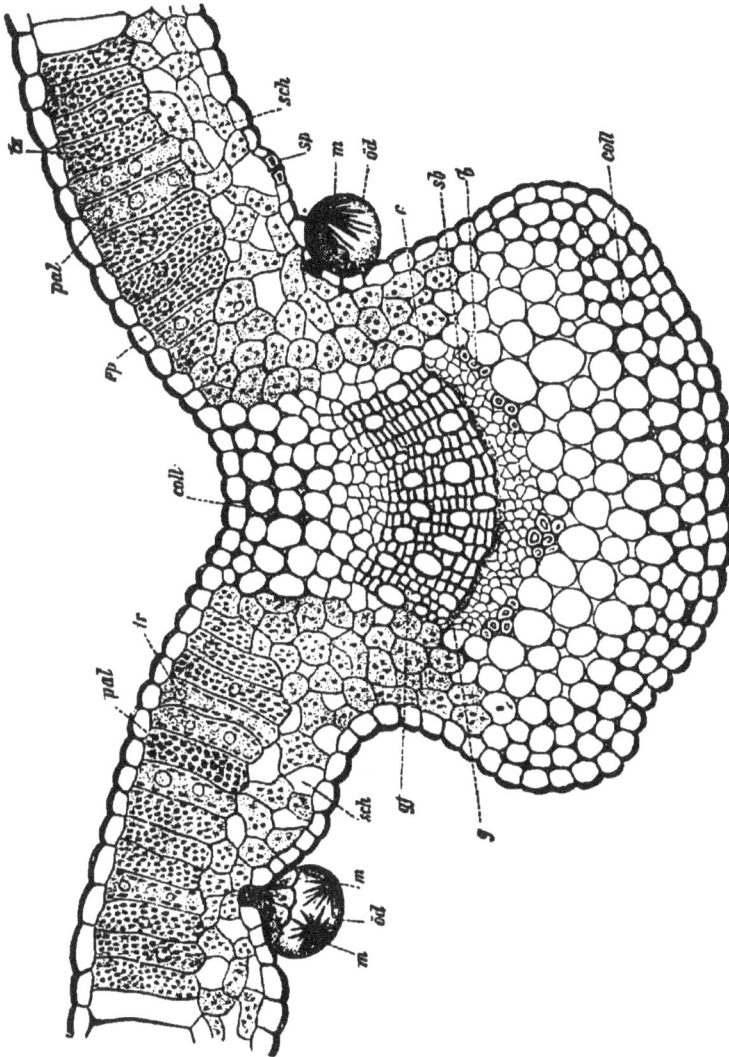

FIG. 129.—Transverse section through a leaf of *Mentha piperita*, at the middle nerve (N); *gf*, vascular bundle; *c*, cambium; *sb*, sieve portion (phloëm); *g*, vascular portion (xylem); *b*, bast-cells; *coll*, collenchyma; *ep*, epidermis; *pal*, palisade tissue; *sch*, spongy parenchyma, both, but especially the former, filled with chlorophyll granules and provided with drops of oil (*tr*); *öd*, oil gland, with drops of volatile oil and crystals of menthol (*m*); *sp*, stoma (Tschirch).

Adolf Meyer,[1] as also Lemaire (see page 51), have utilized

[1] "Anatomische Charakteristik officineller Blätter und Kräuter."
Abhandl. der naturforschenden Ges., Halle, xv. (1882).

the anatomy of leaves, especially the epidermis and the tri-chomes, for the purposes of diagnosis.

5. *The Conducting System.*

When a leaf of the *plantain* (*Plantago*) is torn off or a *maize* stem is broken, there project from the fractured surface numerous fine threads. If the fibrous, fractured place is evenly cut with a sharp knife, it may be seen, even with the unaided eye, that there is a large number of compact, isolated dots imbedded in a more delicate tissue. If the *maize* stem is exposed to decay, only a bundle of very long, fibrous threads finally remains, sur-roundered by a delicate membrane, the cuticle. These threads, as is shown by an anatomical comparison, correspond to the dots upon the transverse section. The threads are termed *fibro-vascular* [1] *bundles, vascular bundles,* or *conducting bundles.* As is already evident from their considerable length, they serve primarily for the conduction of substances, chiefly in the longitudinal direction of the organ.

The same extended threads we meet with in the *maize* leaf. If the latter (or any elongated leaf of a monocotyledonous plant which may be chosen) is held toward the light, a large number of nearly parallel, lighter colored threads (nerves) may be seen in the green tissue.

The nerves do not appear so regular in a dicotyledonous leaf. Here they are variously branched, anastomose with each other, and form a delicate network of fine lines. This is rendered prominent, in an especially handsome manner, when leaves like those of *Digitalis* (Fig. 130), *Datura* or *Matico* are rendered transparent by long maceration in alcohol (of about the specific gravity 0.900), or when freed by decay from the parenchymatous fundamental tissue [2] (Ettingshausen's leaf-skeleton).

That which applies to the leaf and stem is also applicable to the roots. Upon a transverse section of *Rhizoma Filicis*, for example, may be observed a double circle of such threads

---

[1] *Fibra*, fibre, fibre-shaped cell, and *vas*, vessel.

[2] Fundamental tissue in the sense explained on page 176.

or bundles (Fig. 131, f), which likewise represent the frame-
work of the root-stock when the remaining tissues have been

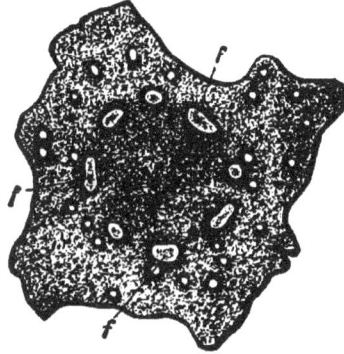

Fig. 131.

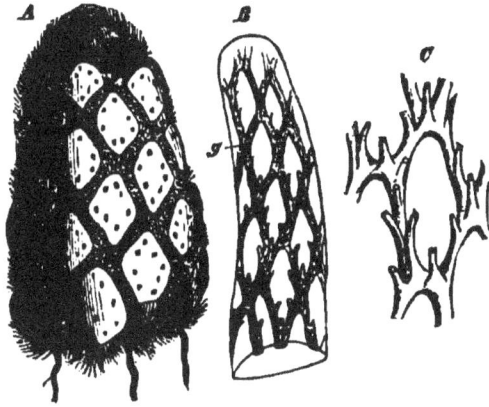

Fig. 120.   Fig. 132.

Fig. 130.—Leaf of *Digitalis purpurea* (Planchon).

Fig. 131.—Transverse section through the underground stem of *Aspidium Filix mas;*
f, threads or vascular bundles (Berg).

Fig. 132.—*Rhizoma Filicis maris* (Sachs). *A*, Front end of the rhizome, showing in the
bright, rhombic spaces the places where the threads enter the leaf-bases (which are here
cut off). *B*, A decomposed piece of the rhizome, g threads or vascular bundles. *C*, A
more highly magnified portion of a bundle.

removed. If such a root-stock be placed in a liquid prepared from decomposed meat, after a short time the parenchyma will be destroyed, and, after washing away the remnants of the same, the far more resistant threads alone remain behind. These do not run parallel in this case, but are variously intertwined (Fig. 132).

In the root-stocks of the Zingiberaceæ and in *Rhizoma Caricis* there are numerous isolated bundles distributed through the fundamental tissue; in *Sarsaparilla* and in *Rhizoma Graminis* they are brought together in the form of a ring (vascular-bundle ring). Differently constructed from these (see below), but similar in their entirety, are the vascular bundles in dicotyledonous roots, as likewise in the dicotyledonous stem, which unite to form a continuous "ring." Dicotyledonous roots often possess only a central, axial bundle (*Ipecacuanha, Taraxacum, Levisticum,* rootlets of *Arnica, Valerian,* and *Helleborus ;* compare also the Figures 119, 120, 122).

But we meet these bundles also elsewhere on every hand. The fruit-pulp of *tamarinds* is traversed by such coarse, string-like, vascular bundles, and the shell of the *almond* is covered with them. They occur in the arillus of the *Myristica* (*Macis*), as well as in the mericarps of the *Umbelliferæ,* in the calyx of the *clove,* as well as in the stigma of the *Crocus*—everywhere forming long threads, which serve for conveying and for conducting away organic and inorganic building material.

The elements of the vascular bundles are so constructed that the impediments to movement are restricted to a minimum. The transverse walls are greatly reduced, often lacking altogether at wide intervals, or, when present, provided with pores [1] (vessels), or even pierced with holes (sieve-tubes), thus having the diffusing surfaces greatly enlarged.

Of what elements is such a **vascular bundle** or **conducting**

[1] The term pores, or pits, is applied to all those thin places of the membrane which are still closed only by the middle lamella. By age the pits occasionally become actual perforations (as in the wood of the *Coniferæ*); compare page 153.

bundle[1] composed ? In the first place, it is necessary to exclude therefrom any bast-cells (Fig. 133), which are, however, not at all regularly united therewith even in the monocotyledons.[2] So long as it was not yet known that the bast-borders of the vascular bundles perform exclusively mechanical functions, they could be considered, and quite properly so, from purely anatomi-

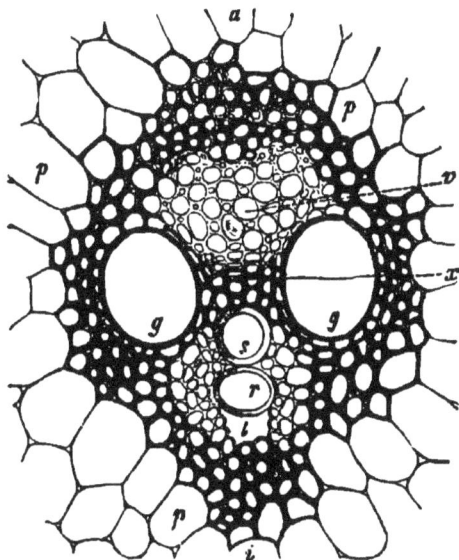 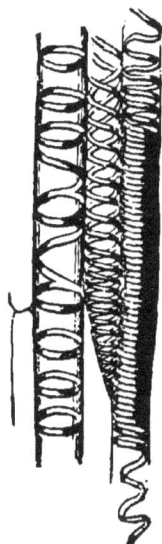

Fig. 133.   Fig. 134.

Fig. 133.—Transverse section through a collateral bundle of the *maize* stem (monocotyledonous type). *a*, outer; *i*, inner; *p*, fundamental tissue; *r*, annular vessel; *s*, spiral vessel; *g*, pitted vessel; *l*, intercellular space (containing air); *x*, wood-cells; *v*, phloëm. The entire bundle is surrounded by a sheath of bast-cells (Sachs).

Fig. 134.—Spirals capable of uncoiling and annular vessel from *Bulbus Scillæ* (longitudinal section).

cal reasons, as belonging to the latter. Since it is known, however, that all bast-cells only serve to impart firmness to the plant,

[1] With regard to the structure of the vascular bundles, which can only be briefly mentioned here, compare De Bary, " Anatomie," p. 328 *et seq.*

[2] The bast border of the cribrose or sieve portion has been termed bast, that of the vascular portion, libriform. Compare the table on page 217.

and have nothing to do with the conduction of substances, they must also be separated from the conducting threads or bundles.

A typical vascular bundle (Fig. 133) consists accordingly only of the *vascular portion* (Fig. 133, *x r s g*, xylem [1] in part) and the *cribrose* or *sieve portion* (Fig. 133, *v*, phloëm [2] in part). If the terms, xylem and phloëm, are so comprehended that the bast and libriform cells are not regarded as necessary constituents of it, the "term sieveportion" is synonymous with that of phloëm, and "vascular portion" with that of xylem. [3]

The **vascular portion** consists of *vessels, tracheïds,* and *wood-parenchyma.*

The **vessels** (*tracheæ*) are very long tubes, which often traverse uninterruptedly the entire plant organ, and the walls of which, in consequence of unequal growth in thickness, are variously thickened (annular vessels, Fig. 134, 133 *r*, spiral vessels, Fig. 134, 133 *s*, trabecular and scalariform vessels, Fig. 135 *fr*, 133 *g*, 142, pitted vessels). They are formed by the disappearance of the transverse walls in a row of cells (fusion of cells). In coniferous wood, the vessels, as a rule, are wanting (*Lignum Juniperi*).

[1] Ξυλον wood.
[2] Φλοιον bark of a tree.
[3] For the sake of clearness these terms may be arranged in a table :

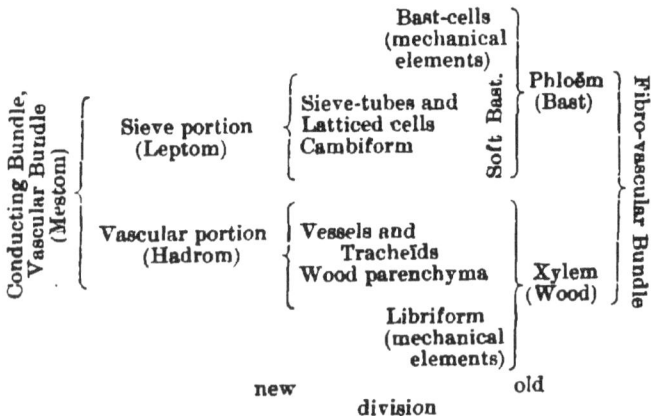

Conducting Bundle, Vascular Bundle (Mestom)		Sieve portion (Leptom)	Sieve-tubes and Latticed cells Cambiform	Soft Bast.	Phloëm (Bast)	Fibro-vascular Bundle
			Baast-cells (mechanical elements)			
		Vascular portion (Hadrom)	Vessels and Tracheïds Wood parenchyma		Xylem (Wood)	
			Libriform (mechanical elements)			

new   old
division

The **tracheïds** (wood-cells) are prosenchymatous and likewise
elongated cells, but these are far from attaining the length of the
vessels. Their walls are mostly pitted. In coniferous wood,
which, as a rule, consists only of tracheïds, areolated or bordered
pits occur (Figs. 64, 72, 136).
The last terminations of the nerves in leaves consist only of
tracheïds. The vessels and tracheïds convey air and water; they
form the conducting system for water and the inorganic nutri-
ment.
The **wood-parenchyma** is not necessarily present, but is de-

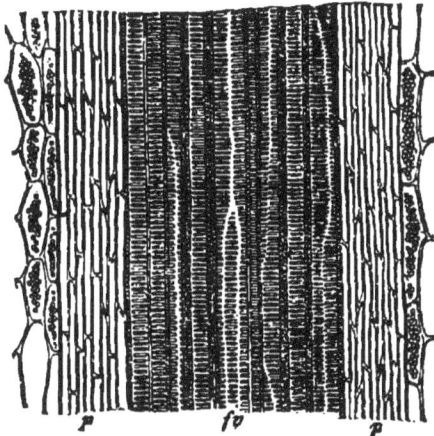

Fig. 185 —Vessels (*fv*) from *Rhizoma Filicis*, thickened in a scalariform manner
(Berg).

veloped in many cases. In contradistinction to the prosenchymatous, tracheal tissue, the former consists of thin-walled cells,
which are connected with the neighboring vessels and cells of the
medullary rays by means of pores, and are filled with various
substances (protoplasm, starch, oxalate and tannin). The wood-
parenchyma serves for conducting the carbohydrates and for the
storage of starch, and either traverses the vascular portion in
the form of more or less isolated cells or occurs in the form of
bands (*Lignum Campechianum, Fernambuci, Guaiaci, Sandali*).

The very dense wood of *Lignum Quassiæ* (Fig. 137) contains groups of parenchyma (*p*), which run transversly in the form of bands from one medullary ray to the other, and intersect the

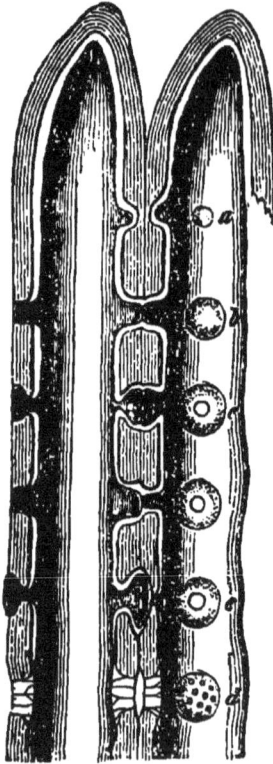 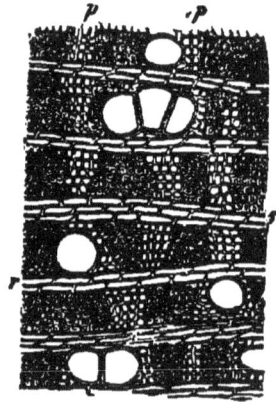

Fɪɢ. 136.   Fɪɢ. 137.

Fɪɢ. 136.—Schematic representation of a longitudinal section of two tracheids from coniferous wood, with the different forms of pitting of the membrane (*a–d*), viewed transversely and from the surface (Hartig).

Fɪɢ. 137.—Transverse section through *Lignum Quassiæ jamaicense;* the medullary rays, which form two or three rows of cells (*r*), are connected by transverse bands of wood-parenchyma (*p*); between them is the libriform tissue.

wood in the form of numerous, closely arranged concentric rings.[1] By their wider and not thickened cells they present a

[1] Compare also Berg's " Atlas," plate xxvii., Fig. 65 (*Lignum Campechianum*), plate xxvii., Fig. 64 (*Lign. Guaiaci*).

very sharp contrast to the zones which surround the large vessels, so that the varieties of wood just mentioned and others
derived from tropical trees appear as if they had *annual rings*.
A transverse section, however, soon teaches that these *bands of
parenchyma* do not form connected rings, and that they also differ
in other respects essentially from the annual rings. A longitudinal
section shows still more distinctly that they consist of parenchymatous tissue. They are designated as *apparent rings*.[1]

Such groups of wood-parenchyma never occur in the vascular
portions of monocotyledons and in the young wood of dicotyledons, but they are not rare in the older wood, the general structure of which may here be considered.

The most marked distinction between the monocotyledonous
and the dicotyledonous stem depends upon the development of
a cambium in the latter (Figs. 129, 138).

The **cambium** (cambium ring, thickening ring) belongs to
the formative tissues, and effects by its activity the *growth in
thickness*. Consequently, where a cambium is wanting, as in
the monocotyledons, a (secondary) growth in thickness is also
excluded. The stems, therefore, remain slender and thin (*Palms,
Bambusa*[2]). The isolated bundles found in monocotyledons
become initially formed as such already at the growing point,
and no merismatic layer is maintained between the vascular and
sieve portions (compare Fig. 133).

It is otherwise with the dicotyledons and the gymnosperms.
Here there appears between the interior, vascular portion, and
the exterior, sieve portion, a layer consisting of very thin-walled,[3]

[1] The fine transverse undulations or horizontal stripings which are
observed upon tangential sections of many woods (*Picrasma excelsa,
Pterocarpus santalinus, Guaiacum, Cæsalpinia, Diospyros*, and others)
proceed, according to Von Höhnel (*Berichte d. deutsch. bot. Ges.*, ii., 3),
from the horizontal rows of equally large medullary rays or from a trabecular arrangement of the pits of the tracheïds (*Tamarindus*), or from both
causes at the same time.

[2] Only some tree-like Liliaceæ (*Aloë, Dracæna, Yucca*) possess a secondary growth in thickness.

[3] In drugs the cambium cells are, therefore, either torn, or at least
greatly distorted and bent. Upon a transverse section of the stems and

tabular cells, rich in protoplasm, which are in a state of active, tangential division. By the activity of this layer vessels are separated on the inner side and sieve elements on the outer (Fig. 138).

Since this activity is not manifested throughout the entire year with the same degree of productiveness, but is more energetic in the spring than in autumn, the elements of the spring wood are more numerous than those of the autumn wood.[1] Since in winter the activity of the cambium ceases completely, it follows that the large spring cells must directly follow upon the narrow autumn cells. There is hereby produced, upon

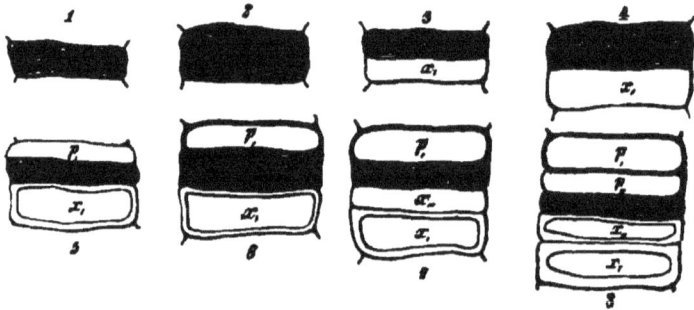

FIG. 138.—Schematic representation of the activity of a cambium-cell (c). 1. before the beginning of its activity; 2. the cell has become radially extended, and (3) divided: a xylem cell (x') has been separated. In 4 the cambium-cell is again extended, while the xylem cell has already become thickened. 5. the cambium-cell by repeated tangential division has this time separated a phloëm cell (p'). While the first-formed xylem cell (x') becomes further thickened, the cambium cell is again extended (6), again a tangential wall appears within it, and the second xylem cell (x'') and soon afterwards (8) the second phloëm cell (p'') is separated. The former become strongly thickened, the latter remain thin-walled (Tschirch).

a transverse section, a visible delineation of rings, for the most part concentric, which are designated by the name of *annual rings*

roots of dicotyledons, the cambium zone frequently appears as a circular line, distinguished by a darker color (*Radix Glycyrrhizæ, Rad. Calumbæ, Rhiz. Rhei, Stipites Dulcamaræ*).

[1] That the cells become narrower and smaller toward autumn, does not proceed (as Sachs, De Vries, and others have stated) from an increased pressure of the bark in autumn.

(Fig. 180 *jg*). In the case of dicotyledonous foliage trees, the distinction between autumn and spring wood is also further in creased by the fact that the latter is much richer in vessels.[1] (Fig. 78 *jf*).

The so-called wood-ring of dicotyledonous stems is produced

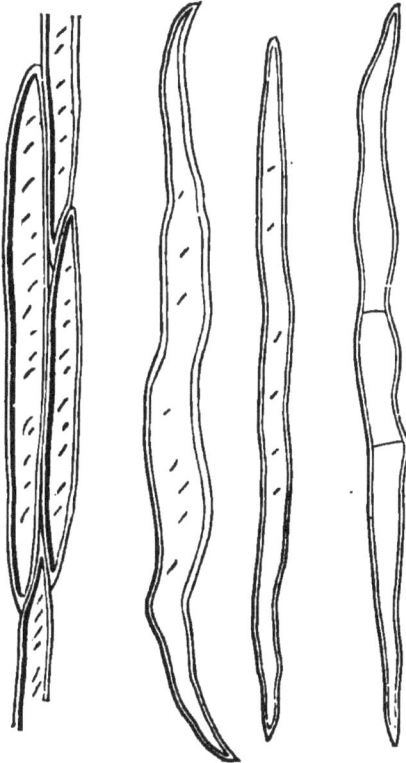

Fig. 139.—Libriform from *Quassia* wood.

by the confluence of isolated vascular bundles, in their vascular as well as in their sieve portions, through the activity of an (intrafascicular) cambium, the bundles being originally formed in a loose circle.

[1] Berg's " Atlas," plate xxv., 60, and plate v., 21.

The elementary organs of the *wood*,[1] as the vascular portion of stems may be briefly termed, are as follows: the *vessels* and *tracheïds*, or tubes conveying water, the *wood-parenchyma cells*, which serve for conducting the carbo-hydrates, and the *libriform cells*, or the specifically mechanical elements of the wood. Besides these, the wood is traversed in a radial direction by the medullary rays.

The first three forms of cells have already been considered.

The **libriform cells**[2] (wood-cells, wood-fibres) are the bast-cells of the wood (Figs. 137, 139), and therefore, strictly considered, belong to the mechanical system (see the latter). They are of service, however, especially in the transitional forms, to the other elements of the wood, occasionally also for conducting and storing nutritive material, and are therefore sometimes provided with contents. They are prosenchymatous, thick-walled, provided with cleft-like, oblique pores, and are never as long as the true bast-cells.

The **medullary rays** (parenchyma rays) are aggregates of cells consisting of one or several rows, always built up of cells which are very much extended radially (Figs. 78, 137, 144, 149), and which pass from the medulla through the cambium and the sieve portion, often penetrating deeply into the bark (Figs. 100 *r*, 118 *m*, 149, 137 *r*, 180, 181).[3]

When the interior of stems is filled with fundamental tissue, the so-called *medulla*, it is thus, by means of these radiating lines of cells, placed in connection with the tissue of the outer bark, which is located on the periphery of the fibro-vascular bundles. These rows of cells are therefore very appropriately called *medullary rays*. Their development in a vertical direction is governed by the number of rows of cells placed over each

[1] The anatomy of the wood has been diagnostically utilized by Nördlinger ("Anatomische Merkmale deutscher Wald- und Garten-holzarten," Stuttgart, 1881).

[2] *Liber*, bast, *forma*, form. Compare Sanio, "Vergleichende Untersuchungen über die Elementarorgane des Holzkörpers," Bot. Zeit., 1863, 101.

[3] Compare also Berg's " Atlas," xxxvi. to xl., Figs. 86–94 *r*.

other in a compartment-like form.   Very frequently this number is not considerable, so that the medullary ray, upon a trans-

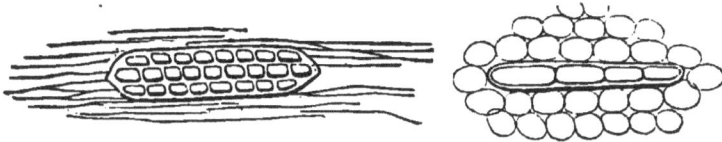

Fig. 140.—Cells of the medullary rays, consisting of one and three rows, cut transversely; tangential longitudinal section of a dicotyledonous stem.

verse section, which is made vertical to its surface of length (tangential to the surface of the stem), represents a cleft filled with

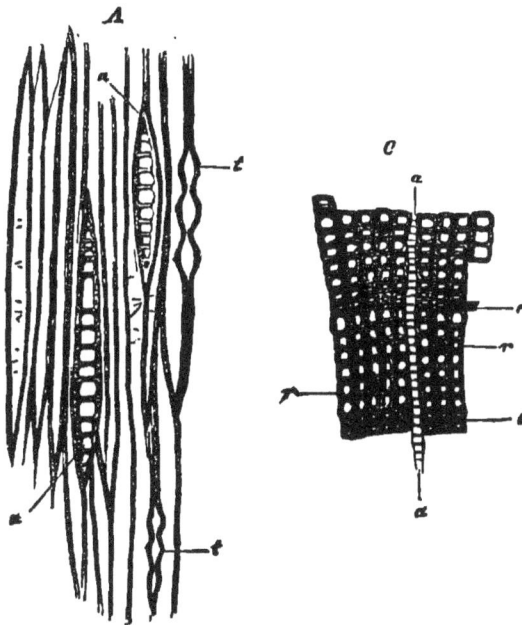

Fig. 141.—Medullary rays from *Lignum Juniperi*, consisting of a single row of
cells (a).
A. Tangential longitudinal section, *a* medullary ray.
C. Transverse section through the wood of the root, *a* medullary ray; *c*, annual rings.

parenchyma (Fig. 140).   In breadth the medullary rays sometimes present but a single row of cells, as, for instance, in *Lignum*

Juniperi (Fig. 141), *L. Guaiaci, L. Quassiæ surinamense,*[1] sometimes two or three rows, as in *Lign. Quassiæ jamaicense* (Fig. 137), and sometimes still more, as in *rhubarb* (Fig. 142). It follows from this that, on a transverse section, the fibro-vascular bundles must appear either in radial rows, or separated

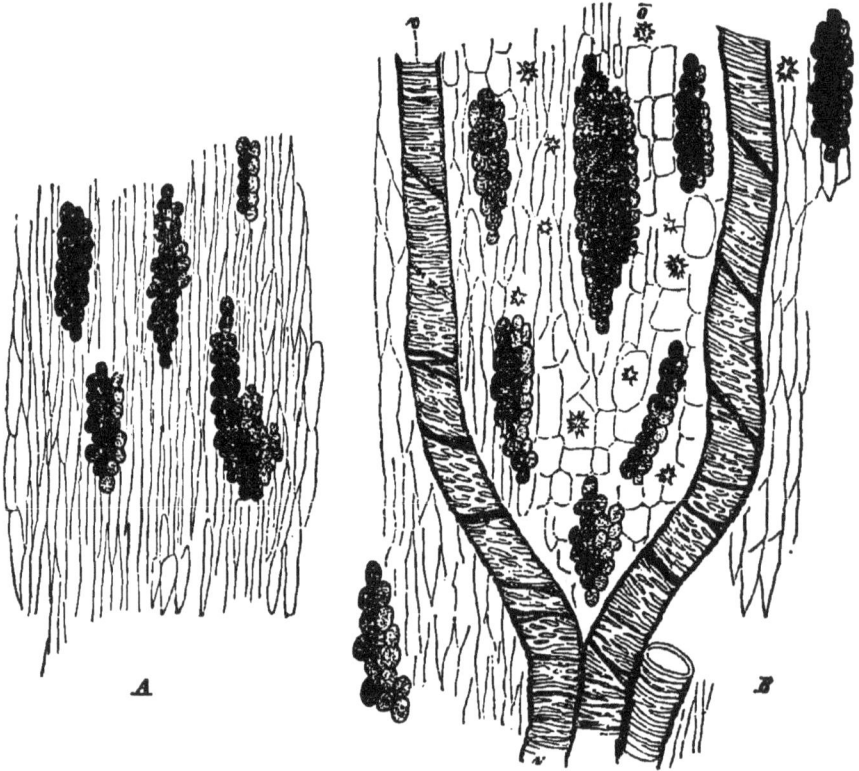

FIG. 142.—Tangential longitudinal section of *Rhubarb*. *A*, Fundamental tissue of the bark with five medullary rays, which are cut transversely. *B*, Section from the xylem portion, which is traversed by netted vessels (*v*) of considerable size; *δ*, rosettes of oxalate crystals.

from each other by large medullary rays. The tissue of the medullary rays, at least within the compass of the fibro-vascular

[1] Berg's "Atlas," plates xxv. to xxviii.

15

system, consists almost entirely of cubical or horizontally (radially) extended (parellelopipedal), thin-walled cells which are joined together in a wall-like form, without intervening spaces

FIG. 143.—Wall-like cells of the medullary rays from *Lignum Juniperi*, on a radial longitudinal section.

(Fig. 143). This regularity is lost at the place where the medullary rays pass into the bark.

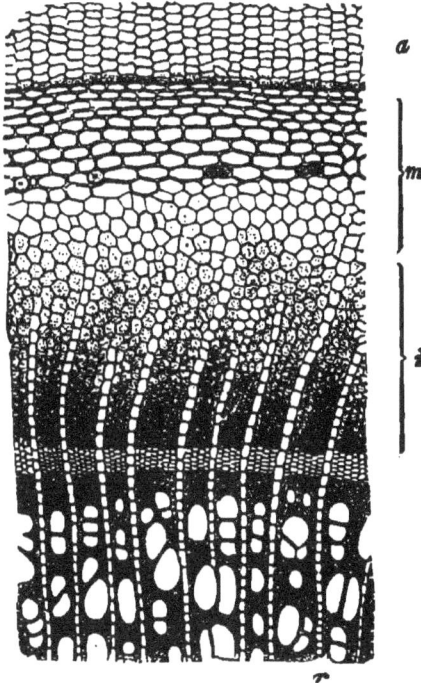

FIG. 144.—Single-rowed medullary rays, *r*, in *Stipes Dulcamaræ*, which gradually become lost in the bark ; *i*, inner bark ; *m*, middle bark ; *a*, outer bark.(Berg).

In the cambial zone, when the growth in thickness is already

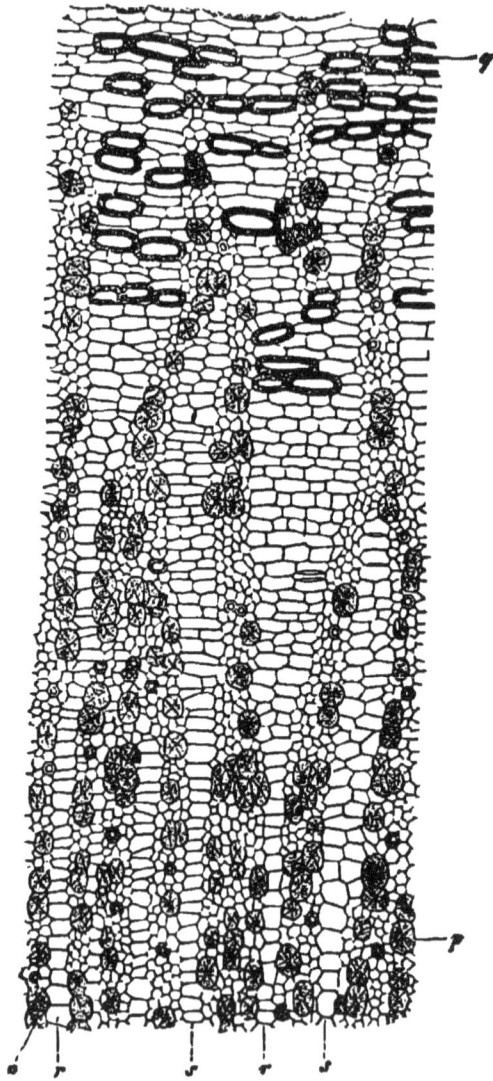

FIG. 145.—Transverse section through phloëm portion of the bark of *Cinchona lancifolia* ; *r*, primary medullary rays; *s*, secondary medullary rays; *q*, stone-cells; *p*, bast cells (Berg).

quite advanced, medullary rays of smaller size are often subsequently developed, which are designated as *secondary medullary
rays*.   They are met with frequently, for example in the *Cinchona barks* (Fig. 145), and, according as they are located in
the xylem or the phloëm, are termed *secondary* xylem (or
medullary) rays, or bark rays.

The great variety of special features in the structure of the

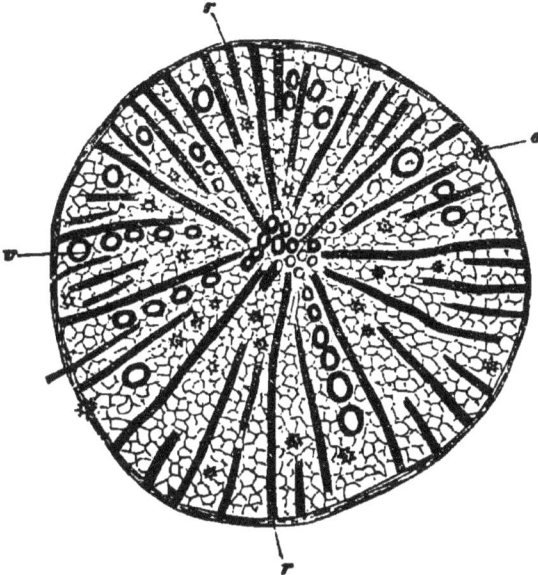

FIG. 146 a.—Transverse section through the peeled root of *Rheum Rhaponticum*.
r, brownish-red medullary rays: v, vessels; ô, crystals of oxalate.

medullary rays affords very noteworthy and characteristic distinctions for many drugs.[1]   As against roots which have a decidedly radiate structure (for instance, *Rad. Rhapontici*, Fig.
146 a), it may suffice here, for example, to refer to such which

[1] In the Coniferæ, as Essner has shown (Abhandl. d. naturforsch.
Ges., Halle, 1882), the number and height of the medullary rays, as well
as the form and size of their cells, possess no diagnostical value.

have no distinctly marked medullary rays, either in the wood
or in the bark (*Radix Ipecacuanhæ* and in *Rad. Taraxaci*).

The medullary rays of *rhubarb* (Fig. 146 *b*), which only occur regularly in the cortical portion, are very characteristic
with regard to their course and contents. Within they are
variously serpentine, and therefore, since they consist of cells
with reddish-yellow contents, produce the peculiar spots on the
surface of the peeled rhizome and the marbled appearance of
the interior.[1]

In those fibro-vascular bundles where, by the formation of
the bundles themselves, the activity of the cambium is termi-
nated already at the growing point, no medullary rays are

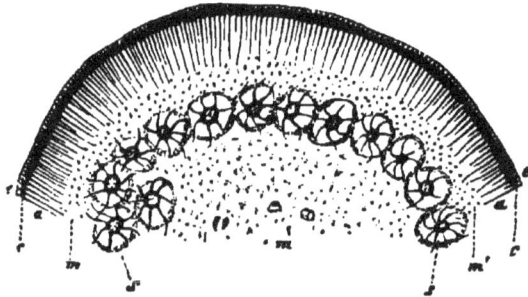

FIG. 146 *b*.—Transverse section of *Rhubarb*. *e*, remnants of the bark, which has
been pared off : *c*, cambium, *a*, medullary rays of the marginal portion; *s*, stellate
spots; *m*, fundamental tissue.

formed; hence they are wanting in the monocotyledons and in
the vascular cryptogams.

Although, for example, the vascular bundles in a number of
monocotyledonous roots are closely connected with each other,
and are arranged in the form of compact, closed rings (*Rhiz.
Graminis*, *Rhiz. Caricis*, *Rad. Sarsaparillæ*, and rootlets of
Veratrum), the medullary rays are nevertheless wanting in these
rings.

The cells of the medullary rays have the function of conducting in a radial direction. But they are also receptacles for

[1] Berg's " Atlas," Plate xii.

reserve substances. Starch is, therefore, often found in them, especially in winter (*Quercus*).

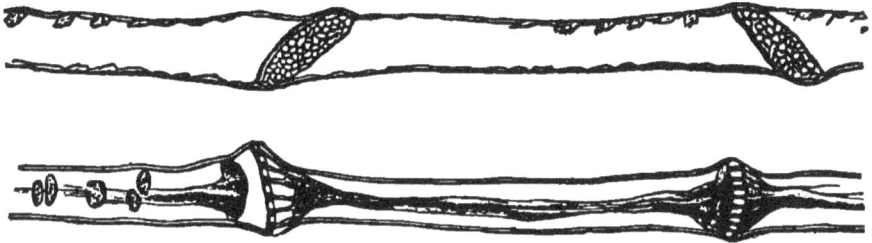

Fig. 147.—Sieve-tubes from *Fructus Papaveris* (Dippel).

In older stems and branches, the **heart-wood** (duramen) and **sap-wood** (alburnum) may be distinguished. The heart-wood,

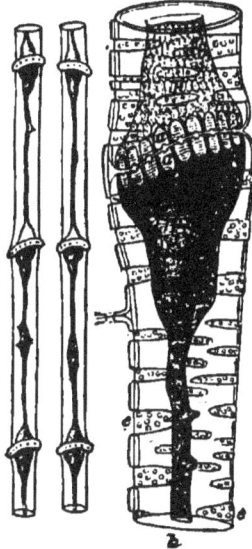

Fig. 148.—Sieve-tubes of *Acer.* (a) more feebly, (b) more highly magnified. a, sieve-plate on a transverse section; e. the same as seen from the surface; p, the contracted plasmatic contents of the sieve-tubes (Hartig).

which is mostly characterized by a darker color (*logwood, red-wood,* and other varieties of *dye-woods, Guaiac*), consists of the older layers of wood, which are no longer vitally active, while the sap-wood is formed of the younger wood, which often comprises but a few annual rings.

The second component of a vascular bundle is the **sieve portion**[1] (phloëm in part). This consists of sieve-tubes (with latticed cells) and cambiform cells ; its elements never become lignified, which, on the contrary, occurs regularly with the elements of the wood.

The **sieve tubes**[2] are elongated, thin-walled cells, which correspond, in their entire structure, to the vessels (in the xylem) (Figs. 147, 148). They serve for conducting non-diffusible, albuminous substances, which they abundantly contain, and are, therefore, mostly divided by oblique, transverse plates, which appear pierced by numerous holes (sieve-plates, Fig. 148 *n*).

The sieve-tubes are often accompanied by small latticed cells, which, from the manner of their development, belong to them.

The **cambiform cells** are elongated, thin-walled cells, very rich in protoplasm, without pits, which are very similar in shape to the cambium cells, and conduct the more readily diffusible substances.

Other parenchymatous elements besides those already mentioned also participate in the conduction of substances, thus, particularly, the **starch sheaths** and **sugar sheaths,** which surround the vascular bundles.

In the dicotyledons and gymnosperms, the medullary rays, as a rule, also intersect the sieve portion (bark rays or phloëm rays); often, however, they remain imperceptible in the tissue of the bark (*Radix Ipecacuanhæ*).[3] If they are distinctly

---

[1] Also incorrectly termed " soft bast." The term " bast " should remain confined to the mechanical elements of bast-cells.

[2] Wilhelm, "Beiträge zur Kenntniss des Siebröhrenapparates dicotyler Pflanzen," Leipzig, 1880. Janczewski, " Vergleichende Untersuchungen über die Siebröhren." Sitzungsber. d. Krakauer Akademie, .881. De Bary, "Anatomie," p. 179.

[3] Arthur Meyer, Archiv der Pharm., 221 (1883), p. 737.

visible, that portion of the **bark** [1] which is traversed by them
is termed the *inner bark* (Fig. 144 *i*, 149 *i*); the portion toward
the exterior which is connected with it, and which mostly con-

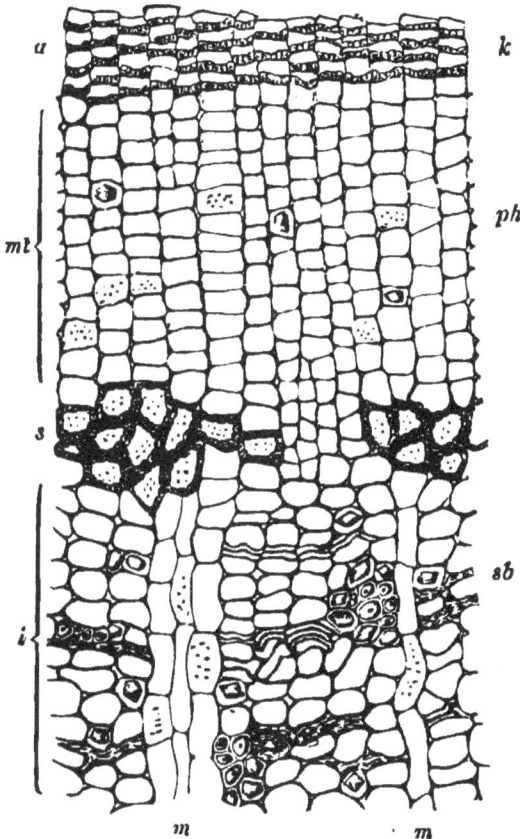

FIG. 149.—Transverse section through the bark of an older stem of *Ceratonia Siliqua*
L. *k*, sclerotized cork-cells: *s*, sclereids: *m*, medullary rays; *sb*, sieve-tubes; *i*, inner
bark; *mt*, middle bark; *a*, outer bark (Moeller).

sists of a mixture of parenchymatous cells containing chlorophyll (bark parenchyma) and of mechanical cells (bast-cells,

[1] With regard to the structure of the bark, compare especially Moeller,
"Die Anatomie der Baumrinden," Berlin, Springer, 1884.

stone-cells, collenchyma), is termed the *middle bark* (green bark, Figs. 144 *m*, 149 *m*); and the epidermis (cork) the *outer bark* (Figs. 144 *a*, 149 *a*).

Naturally, barks are found only in the dicotyledons and gymnosperms. The name is used to designate the entire collection of cells located outside of the cambium. Nevertheless, in the case of monocotyledons the portion of tissue located outside of the nucleus-sheath is also, for the sake of convenience (but incorrectly) termed bark.

In the middle bark, entire lines of cells are often found which are filled with crystals of oxalate. These frequently accompany

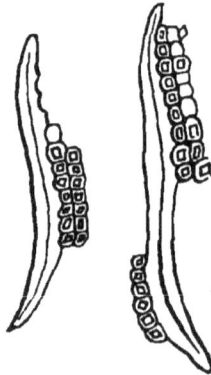

Fig. 150.—Bast-fibres from *Cinchona (China) Alba Payta*, with impressions of the neighbouring crystal cells.

the bast-cells, and are occasionally in such close contact with the latter that depressions are produced. The crystal-cells appear indented into the bast-cells, as in "*Cinchona (China) alba*"[1] (Fig. 150) and in the bark of *Aspidosperma Quebracho*.[2]

If the activity in the entire (cambial) thickening ring progresses in a uniform ratio, then the separated elements of the wood and of the bark are arranged in strictly radial rows.

In the wood, these rows mostly remain preserved (coniferous

[1] Flückiger, Neues Jahrbuch für Pharmacie, xxxvi. (1871), p. 293.
[2] Hansen, " Die Quebracho-Rinde." Berlin, 1880.

wood, *logwood, red-wood*) in consequence of the thickness of the walls of the cells and the absence of dislocating tensions. In the bark, however, especially at later periods, in consequence of the various dislocations, lacerations, and subsequent divisions of the cells, etc., they are seldom distinct (*cinnamon*). Even the bark rays have seldom (for instance, in the *willow*) a precisely radial course (*Cinchonas*). Nevertheless, the radial rows of the elements of the bark may often be followed nearly to the epidermis.

According to the manner in which the distribution of the vascular portion and the sieve portion takes place in the vascular bundle, the following designations are employed. The bundle is spoken of as *concentric* when one of the two parts surrounds the other in a sheath-like manner, that is, either the sieve portion the vascular portion (*Rhiz. Filicis*, Fig. 135), or the vascular portion the sieve portion (*Rhiz. Calami*, Fig. 151, *Rhiz. Iridis*);[1] as *radial*, when the vascular and sieve portions are located radially, one behind the other (the typical structure of the vascular bundles of roots, Figs. 120, 121); as *collateral*, when both parts run beside each other and are in contact laterally (the typical structure of vascular bundles in the stem and leaves of phanerogams).

In the leaf, the sieve portion always lies in a direction toward the lower surface, and the vascular portion always toward the upper surface (Fig. 129 *g, sb*).

If, through the activity of the cambium, the formation of *secondary wood* takes place, the *primary bundles*, which had already previously been formed at the tip without the activity of the cambium, move more and more toward the interior; and in all those cases where they lie relatively widely separated from each other, they project toward the medulla in a bow-like form and constitute the medullary-crown or **medullary sheath.**

The outermost layer of the fibro-vascular body of roots, which remains for a long time capable of development, or, in other words, in a parenchymatous condition, and in which the rootlets become formed, is termed the *pericambium.*

[1] Compare Berg's " Atlas," Plate xxi.. Fig. 51.

6. The Storing System.

Since the plant does not immediately consume all the substances which it takes up or forms, it must possess some arrangement whereby these substances may be stored for future use. The plant therefore develops **receptacles for reserve substances**, in which it deposits the building materials. Such reserve receptacles are the seeds,[1] fruits, rhizomes, tubes, bulbs and perennial roots; indeed, even the stem (especially the medullary rays)[2] can become a receptacle for reserve substances.

The substances deposited in these organs are mostly solid; readily soluble substances (sugar, dextrin) are avoided, as a rule, by the plant. In the *sugar-beet*, however, the sugar is a reserve substance. They are either carbohydrates (starch, inulin, cellulose, for instance in the *Phytelephas* seed), fats (oil) or albuminous substances (gluten, aleurone); but water also is stored for the germinating plant (bulbs),[3] or it is energetically retained by the fully-developed plant (epidermal water-tissue).[4]

7. The Aërating System.

In the development of plants, the interchange of gases plays an important part. The plant consumes carbonic acid (and oxygen) from the atmosphere, and gives up to the latter oxygen (and carbonic acid), as well as aqueous vapor. In order to render possible this interchange of gases, cavities in the tissue are necessary. These collect the gases, and the canals, which serve respectively for their entrance and exit, convey them from without to the interior and inversely.

For these reasons the plant is abundantly provided with **intercellular spaces** (Figs. 129, 127), especially in those parts

[1] The seeds have received appropriate anatomical treatment by Harz, " Landwirthschaftliche Samenkunde." Berlin, 1885.

[2] Malpighi, "Anatomes plantarum idea" (1671), had already recognized the medullary rays as reserve receptacles.

[3] All bulbs (*Bulbus Scillœ*) attract water abundantly, in consequence of the mucilage and sugar which they contain.

[4] Westermaier. Pringheim's Jahrb., xiv., 43.

in which the processes of development are actively effected.
While the cells of the epidermis, of the cambiform, of the cam-

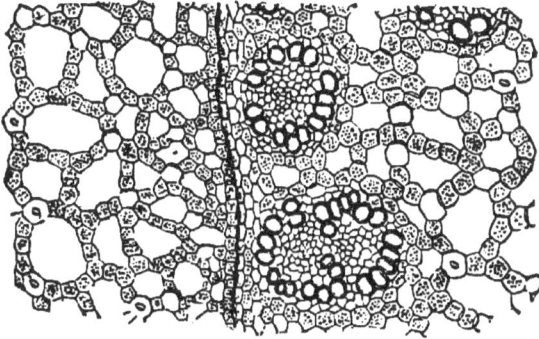

[Fig. 151.—Moniliform tissue from *Rhizoma Calami*. Transverse section through the nucleus-sheath; o, oil-cells.

bium, and of the medullary rays are closely united to each other,
without intervening spaces—the wood, the bark, and the tissue

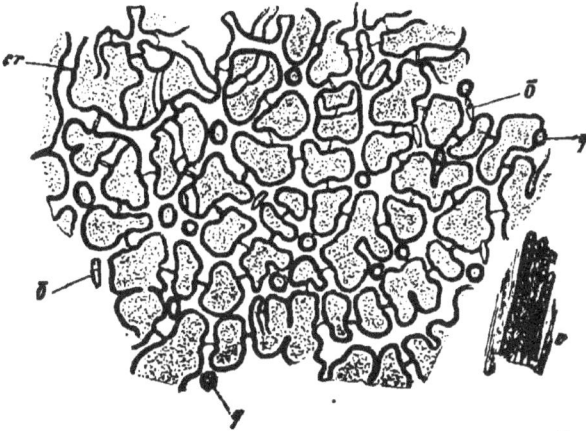

Fig. 152.—Transverse section from the sarcocarp of *Fructus Aurantiorum*; the large, branched cells having sieve-like pores at their points of connection, for example, at *cr*; *q*, branches of cells cut transversely; *b*, imperfectly developed, large crystals of oxalate. The dotted places represent the air-cavities; *v*, fibro-vascular bundle.

of the leaf possess an abundance of such, mostly triangular or
quadrangular, intercellular cavities. Aquatic plants, especially

their immersed portions, also show very large and numerous intercellular spaces (*Rhiz. Calami*, Fig. 151, *Carex* leaves, rhizomes of the *Nymphæa, Carex arenaria* and *Gratiola*[1]). The intercellular spaces produced by the disintegration of cells, as a result of the tensions occurring through growth, may assume the most varied forms, and finally by far exceed the cells themselves in area.

Excellent examples of this are afforded by the thickened margins of *Siliqua dulcis*, the more central tissue of *Fructus*

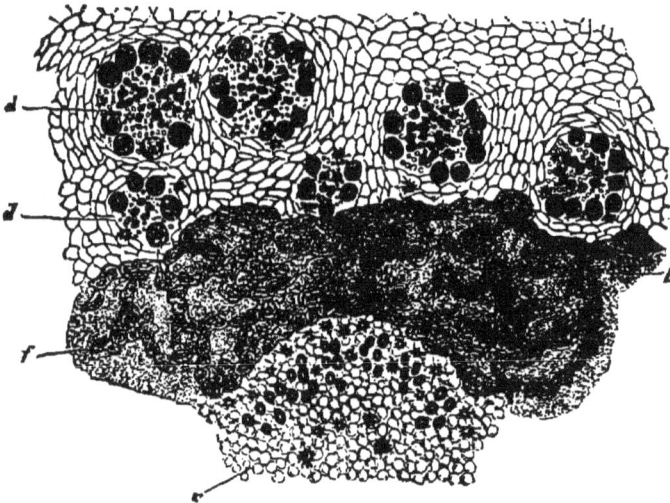

Fig. 153.—Transverse section through *Caryophyllus*, inner tissue; *f*, loose, branched cells, interrupted by large air-cavities; *t, d*, fibro-vascular bundles; *v*, central fibro-vascular bundle.

Aurantiorum (Fig. 152), the rhizome of *Calamus* and a portion of the inner tissue of the clove (Fig. 153).

Finally, wide spaces containing air (air-cavities, lacunæ) may also be found between the tissues, in consequence of the delicately-walled parenchymatous tissue, especially the medulla, not being able in its development to keep pace with the growth in thickness of the other tissues, so that the cells of the former

[1] Berg's "Atlas," plate xxii., Fig. 55.

gradually become destroyed. In this manner the hollow stems (Gramineæ, Umbelliferæ, *Dulcamara*) and roots (*Rhiz. Graminis*) are produced. Some of the air-cavities in many roots and leaves (aquatic grasses) may also be so produced; at least shreds of the cell-membrane are often found on their margins. Lastly, there may also be mentioned the cavities which are formed in many drugs by the ultimate laceration either of lines of cells of the medullary rays (*Radix Bardanæ* seu *Lappæ, Carlinæ*) or of

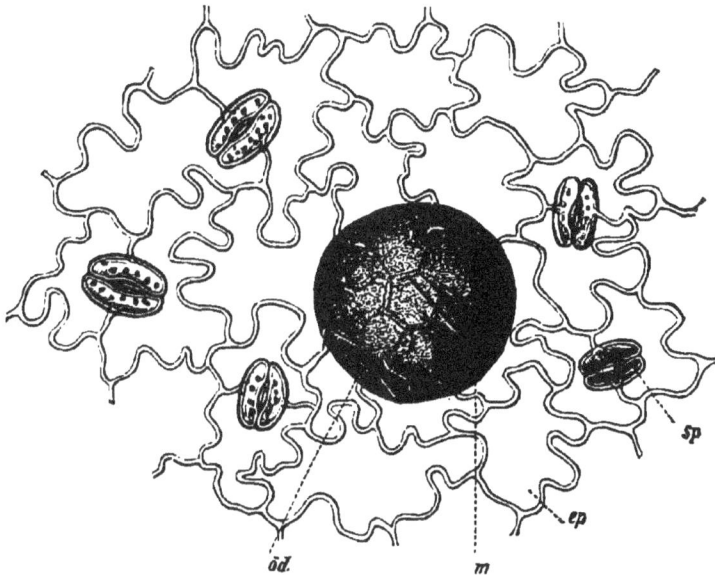

FIG. 154.—Surface view of the epidermis of a leaf of *Mentha piperita; ep*, undulated epidermal cells; *sp*, stoma; *öd*, oil glands (seen from above) with a group of crystals of menthol *m* (Tschirch). Compare also Fig. 129.

the bark (especially of the bark-rays, *Rad. Arnicæ*, Fig. 182 *b*, *R. Levistici, R. Pimpinellæ*).

It is, however, more especially the lower side of leaves which is abundantly aërated (transpirating tissue, Figs. 127, 129). Here are also chiefly found the canals of exit, the **stomata**, which are mostly located at the level of the epidermis or are only slightly depressed below the latter (Figs. 129 *sp*, 155), and which establish a direct connection of the intercellular spaces of the inner tissue with the atmosphere. Besides on leaves, stomata are also found on seed-vessels, young axial organs (stems), and even on the walls of the ovary, but never on roots.

On leaves the stomata occur for the most part only on the lower surface (Figs. 127, 129), or the latter is at least more abundantly provided with them than the upper surface. This applies, however, only to bifacial leaves, the centric leaves (see page 210) having stomata on both sides (Fig. 128). In the monocotyledons the stomata lie mostly in straight lines, the cleft having the same direction in all (Fig. 156); in the dicotyledons they are irregularly distributed (Fig. 154).

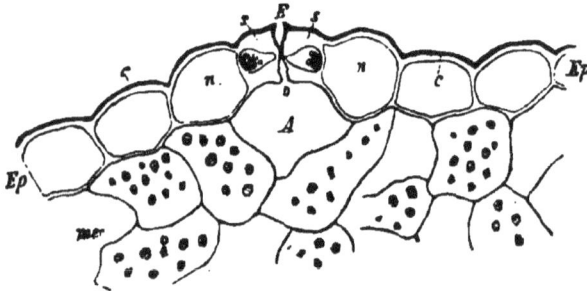

Fig. 155.—Vertical section of a stoma from a leaf of *Mentha piperita;* *Ep*, epidermis: *c*, cuticle; *mer*, leaf merenchyma; *s*, guardian cells; *E*. eisodial opening; *o*. opisthial opening; *A*, respiratory cavity; *n*, lateral cells (Tschirch).

Their form and development [1] is very varied. Plants occurring in dry climates have variously protected (depressed) stomata (Figs. 63, 128).

There are distinguished on the stoma (Fig. 155) the guardian cells (*s*), the central cleft (porus), and the eisodial (*E*), and opisthial opening [2] (*o*). Beneath the guardian cells there is

---

[1] Weiss, Pringsheim's Jahrbücher f. wissensch. Botanik, iv., 1862, p. 425. Strassburger, *Ibid.*, v., plates 34–36. Tschirch, "Ueber einige Beziehungen des anatom. Baues der Assimilationsorgane zu Klima und Standort," Linnæa, 1881. In the latter the *entire literature* relating to the stomata is given up to the year 1881.

[2] Compare Tschirch, loc. cit., p. 140.

always a more or less wide respiratory cavity (Figs. 155 A, 179

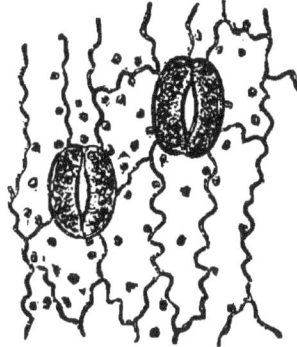

FIG. 156.   FIG. 157.

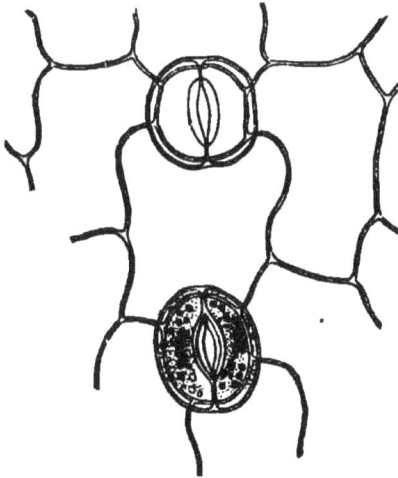

FIG. 158.

FIGS. 156, 157, 158.—Stomata from dicotyledons and monocotyledons on a surface section, as seen from above.

a), and occasionally, beside the guardian cells, peculiarly formed lateral cells (Fig. 155 *n*). The number of the stomata varies exceedingly. Upon a square millimeter of surface of a leaf from ten to eight hundred of them may be present. The distribution of the stomata, as also the shape and size of the epidermal cells, is employed by Bell [1] as a means of distinguishing *tea, elder, willow* and *sloe*, or *black-thorn* leaves. The periderm, like the epidermis, is also occasionally traversed by open gas passages. These are the **lenticels** [2] or bark-pores. They are mostly produced beneath the stomata, and, as they consist of roundish, suberized cells, they form small, sharply circumscribed corky protuberances, which assume on the older organs the function of the stomata, that is, they regulate the interchange of gases. For some barks the lenticels are characteristic (*Rhamnus Frangula, Solanum Dulcamara*). Such corky growths occur also upon leaves (*Eucalyptus*, Fig. 128 *k, Dammara*). [3]

8. *Receptacles for Excretions.*

During the change of matter incidental to the life of the plant, there are formed certain substances which, so far as our knowledge extends, find no further application, and are therefore termed *excretions*. [4]

They may either be produced in more or less peculiarly shaped cells or cellular passages (oil-cells, mucilage-cells, latex-tubes),

[1] " Die Analyse der Nahrungsmittel," i. (Berlin, 1882), p. 36.

[2] Diminutive of the Latin *lens*, a lens, on account of their form. Regarding to the lenticels, compare Stahl, Botan. Zeit., 1873, No. 36. Klebahn, Ber. d. Deutsch. botan. Ges., 1883, and '' Die Rindenporen. Ein Beitrag zur Kenntniss des Baues und der Function der Lenticellen und der analogen Rindenbildungen," Jena, 1884, where the literature is also given.

[3] Bachmann, '' Ueber Korkwucherungen auf Blättern," Pringsheim's Jahrb., xii., p. 190.

[4] Regarding the limitation of the terms excretion and secretion, compare the interesting investigation of Szyszylowicz, '' Ueber die Sekretbehälter der flüchtigen Oele im Pflanzenreich." Denkwürdigkeiten der Krakauer Akademie, 1880 (Bot. Centralbl., viii., 1881, p. 259).

16

or secreted by special cells or aggregations of cells (organs of secretion) in intercellular spaces (oil-passages).

The volatile oils and the resins come primarily under consideration here.

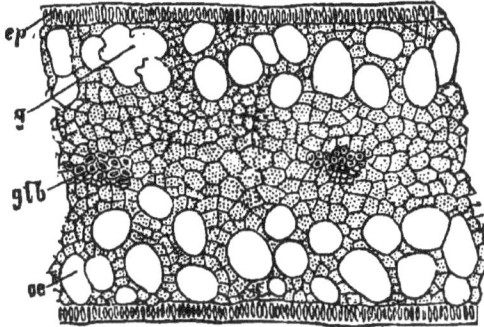

Fɪɢ. 159.—Cross section through the arillus of *Myristica spec.* (*Bombay Mace*); *ep*, epidermis; *gfb*, vascular bundle; *oe*, oil-cells; *g*, a larger oil-space formed through the confluence of several oil-cells.

The latex-tubes were also formerly considered as receptacles for excretions, but it has recently been rendered probable, by

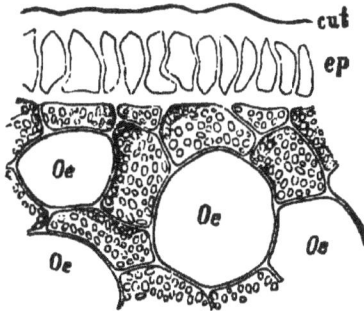

Fɪɢ. 160.—A portion of the same cross section, more highly magnified; *cut*, cuticle (Tschirch). Compare *Pharmac. Zeitung*, 1881, No. 74.

the discovery of points of connection with the assimilating tissue, that they have some share in the transfer of substances, and that their contents are not true excretions, although of more or less importance in the process of development.[1]

[1] In this place reference may also be made to the fact that the latex-

Nevertheless the latex-tubes should be treated of in this place..
Oil-cells. Volatile oil, in roundish or oval cells,[1] is found in
cloves (as much as 20 per cent), in *cubebs* (as much as 13 per
cent), in *mace* (up to 17 per cent, Figs. 85, 159, 160),[2] in the
leaves and barks of the Lauraceæ [3] (*Cinnamomum, Camphora,*
Fig. 161), and in the rhizomes of the Zingiberaceæ (Fig. 125) [4]
and of *Acorus Calamus* (Fig. 151 *a*). Volatile oil, mixed with

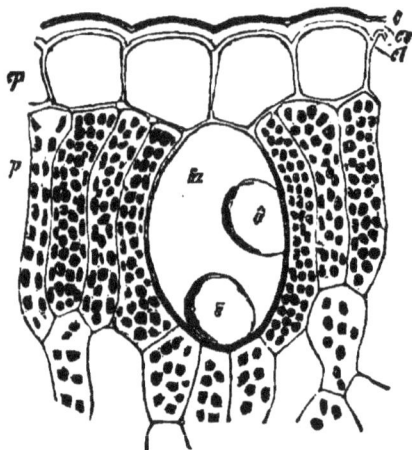

Fig. 161.—Transverse section through a sub-epidermal oil-cell in the surrounding tissue
from a leaf of *Cinnamomum Camphora; ep*, epidermis (*c*, cuticle; *cs*. cuticle layer; *cl*,
cellulose layer), *p*, palisade tissue; *öz*, oil-cell; *ö*, drops of oil (Tschirch).

tubes are frequently connected anatomically with other conducting
organs, especially with sieve-tubes, as, for instance, in *Rad. Taraxaci*
(Figs. 164 and 165).

[1] All the cells dispersed through the fundamental tissue, and differing
in shape or contents (stone-cells, oil-, resin-, mucilage-cells, latex-tubes
etc.), are also designated in general by the name of *idioblasts* (from
ἴδιος peculiar, βλαστός, germ, organic structure). Heinicher (Ber. d.
Deutsch. bot. Ges., ii.) has also recently found in the Papaveraceæ
idioblasts containing *albumen*.

[2] These examples may also be regarded at the same time as the maximum amounts of volatile oil contained in plants.

[3] Berg's "Atlas," Plate xxxvi., Fig. 86.

[4] *Ibid.*, Plate xix.

other substances (coniine), occurs in peculiarly shaped lines of
cells in the fruit of the *Conium*.

The volatile oil fills, or almost entirely fills, the individual
cells. The same applies to the mucilage of the **mucilage-cells** [1]
(see above), such as are met with, for example, in *Rad. Athœæ*,
in the *Orchis* tubers, in *Cortex Cinnamomi*, and in *Cortex
Ulmi*. As a rule, the cells containing oil or mucilage are somewhat larger than the cells of the surrounding tissue (*Orchis,
Cort. Cinnamomi*,[2] Figs. 160, 161, 186 *ch*).

As an example of the occurrence or absence of mucilage-
cavities under special conditions of growth (and therefore by no
means regularly) we may cite *Laminaria hyperborea* Foslie (*L.
Cloustoni* Edmonston).[3]

In addition to volatile oil and mucilage, crystals,[4] tannin [5] and
resin also occur as constituents of true secreting cells. Höhnel [6]
has futhermore shown that kino,[7] from the bark of *Pterocarpus
Marsupium* Roxburgh, also occurs in extended (100–500 μ)
cells or tubes, united to form long bundles, which are not suberized. On the other hand, the oil-tubes of the Indian species
of *Andropogon* (*Andropogon Schœnanthus* L. was examined),
from which we obtain the fragrant volatile "grass-oils," [8] are
suberized.

The previously mentioned **glandular hairs,** the terminal

---

[1] Berg's "Atlas," Plate xxiii., Fig. 57.

[2] *Ibid.*, Plate xxxvi., Fig. 86 *m*.

[3] M. Foslie, "Ueber die Laminarien Norwegens" (Christiania Viden sk.-
Selsk. Forhandl., 1884, No. 14), p. 47 of the reprint.

[4] The true crystals of many plants (especially of monocotyledons), as
well as the cystoliths, are to be regarded as excretions.

[5] We place the tannic matters here under the excretions, although, like
the constituents of the laticiferous juices, they probably still participate in the processes involving the transmutation of matter.

[6] "Ueber die Art des Auftretens einiger vegetabilischer Rohstoffe in
den Stammpflanzen." Sitzungsberichte der Wiener Akademie, 89 (1884),
January number.

[7] To what extent such special substances occurring in the plant are to
be considered as excretions still remains undetermined.

[8] Flückiger, "Pharmakognosie," p. 157.

cell of which contains a secretion, must likewise be classed here. These glandular hairs are either trichome formations (epidermal glands, glandular hairs, tufts, scales, compare Figs. 129 *öd*, 154, 96), and as such have already been treated of, or they are produced by the protuberances of cells of the inner tissue, which border on intercellular spaces (root-stock of *Aspidium Filix mas*).

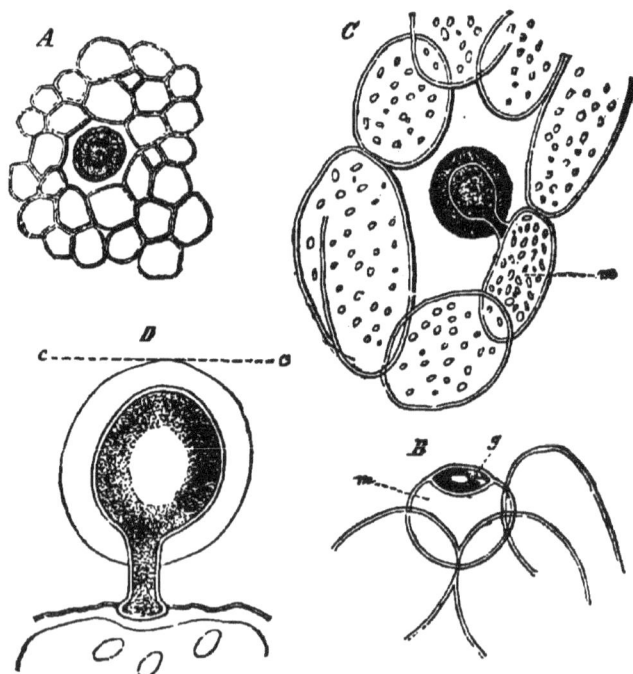

FIG. 162.—From the fundamental tissue of the underground stem of *Aspidium Filix mas*.

A.—Intercellular space from the younger tissue, in the middle showing a gland, as seen from above and covered with a green exudation.

B.—Longitudinal section through a wall-cell (*m*), of the intercellular space, from which the gland (*g*) begins to grow as a protuberance.

C.—Longitudinal section through the mother-cell (*m*) filled with starch, from which the gland, which is borne upon a stalk, projects as a daughter-cell into the intercellular space. The daughter-cell has permitted its green contents to appear upon the surface (as already represented from above, in *A*).

D.—A single gland more highly magnified, and freed from its covering, with the exception of a delicate film (*cc*), by boiling with alcohol (Schacht).

The intercellular spaces in the fundamental tissue of the root-stock and in the leaf parenchyma of *Aspidium Filix mas* (Fig. 162), which have been described by H. Schacht,[1] are very remarkably constructed.   All of the boundary cells of these cavities do not assume a special form, but some few of them grow in a spherical manner into the hollow space, through a protuberance of the delicate wall in one or two places.   The daughter-cell, which is thus produced, becomes immediately bounded by a transverse wall, and elevated upon a small stalk over the mother-cell in a head-like form (the "Zottenkopf," or tufted head, of Hanstein).   These small glandular cells thus remind

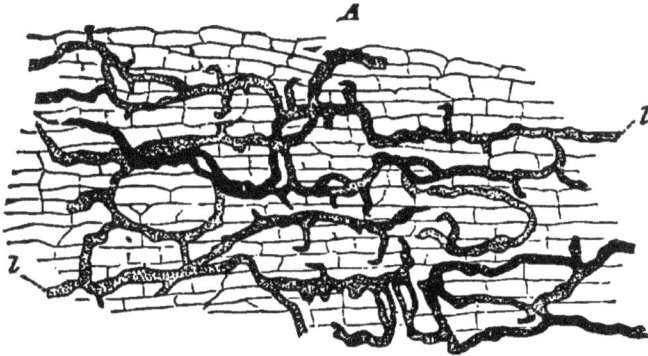

Fig. 163.—Latex-tubes (*l*) from *Radix Taraxaci*.   Tangential longitudinal section through the inner bark.

one of the more simple forms of the above-described oil-producing trichomes of the Labiatæ.   In the fern-root the glands in their terminal head-like cell at first contain protoplasm, in which after a short time greenish oil-drops occur; these are finally forced out upon the surface of the gland, and envelop it as a thin greenish layer.   This section consists for the most part of the peculiar filicic acid, which, by longer preservation under glycerin, crystallizes in long needles; volatile oil is wanting here, or is present in but very slight amounts.   Such intercellular glands have also been met with by one of us (F.) in

[1] Pringsheim's Jahrb. f. wissenschaftl. Bot., iii. (1863), p. 352.

the (non-officinal) root-stock of *Aspidium spinulosum* Swartz; in the other ferns of our region they are wanting.

A second form of secreting cells are the **latex-tubes** or **laticiferous ducts.**[1] In most cases (especially when the tubes are very long) they are produced, like the vessels, by the resorption of the transverse walls of lines of cells lying over each other (cell-fusion). In their simplest initial formation they are distinguished, however, from the neighboring parenchyma cells, like the mucilage-cells and the oil-cells, only by their contents and somewhat more considerable width, as for instance in *jalap*

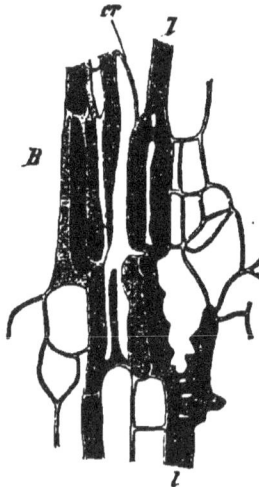

FIG. 164.   FIG. 165.

FIG. 164.—Longitudinal section through the outer latex-zone of *Radix Taraxaci*, more highly magnified; *cr*, sieve-tubes; *l*, latex-tubes.

FIG. 165.—Longitudinal section through one of the inner latex-zones of *Rad. Taraxaci*, in which the tubes (*l*) are accompanied by sieve-tubes (*cr*).

[1] With regard to latex-cells and latex-tubes, compare Meyen, " Die Secretionsorgane der Pflanzen," Berlin, 1837.—Hanstein, " Die Milchsaftgefässe," Berlin, 1864.—Dippel, "Entstehung der Milchsaftgefässe," Rotterdam, 1865.—Vogl, " Beiträge zur Kenntniss der Milchsaftorgane," in Pringsheim's Jahrb., v., 31.—David, " Ueber die Milchzellen der Euphorbien," Breslau, 1872.—E. Schmidt, Botan. Zeit., 1882. Scott, Inaugural Dissertation," Würzburg, 1881, etc.

tubers (Fig. 166) or in the Chinese *galls* (Fig. 91). In the *Cinchona barks*, the latex-tubes are distinguished by their considerable length, often also by a far greater diameter; in other cases, as in *Fructus Papaveris* and in *Caricœ* [1] (Fig. 167), they are developed as branched systems of canals. In this manner abundantly branched latex-tubes traverse definite layers of *Radix Taraxaci* (Figs. 163 to 165). Here the system of these tubes, without regard to the overground portions, is only developed in the bark; in *Lactuca virosa* it extends also to the central parenchyma of the stem, together with all other parts of this plant.

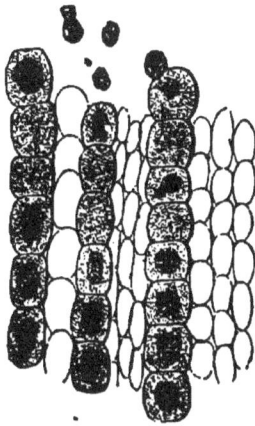

FIG. 166.—Cells from *Tuber Jalapœ* containing laticiferous juice.

One may accordingly speak of *latex-cells* and *latex-vessels* or *tubes*. The former are true cells, and, when formed already in the germinating plant, may often become very long, and, indeed, absolutely branched, without being divided by transverse walls, as for instance in the Urticaceæ, Euphorbiaceæ and Asclepiaceæ. The milk-tubes, on the contrary, are *lysigenic* [2] passages, that is, canals formed by the solution of the trans-

---

[1] The latex-tubes of the fig are so striking that by means of them one may easily recognize an adulteration of coffee with "fig coffee" (roasted figs).

[2] *Δύω* a dissolution, and *γεννάω* I produce.

verse walls. With these are to be classed the tubes of the Cichoriaceæ (*Taraxacum*) and those of the Papaveraceæ and Campanulaceæ.

If a resorption of the transverse walls does not take place, but the short latex-cells lie in rows over each other, lactescent lines of cells are produced, as for instance, in the Convolvulaceæ (*Tuber Jalapœ*, Fig. 166); *gutta percha* (mostly from *Dichopsis Gutta* Bentham et Hooker) also occurs in such lines of cells containing laticiferous juice.

The contents of the latex-tubes is a mixture of very variable composition.[1] Besides the more commonly distributed substances,

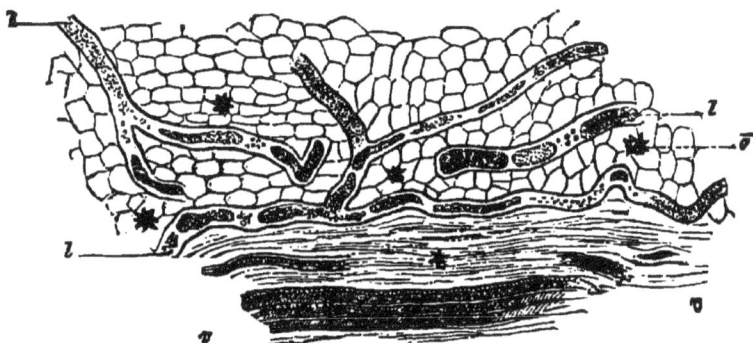

FIG. 167.—Latex-tubes of the fig; tangential section through the more central layer. *l*, tubes; *ð*, rosette of crystals of calcium oxalate; *v*, vascular bundle.

such as salts (calcium malate), starch and protein bodies, very many, if not all, laticiferous juices contain caoutchouc. Peculiar bitter principles and alkaloids (as in *opium*) also occur in them, and some laticiferous juices are therefore medicinally valuable (*opium, euphorbium, lactucarium*). In the latex of species of *Euphorbia* is found the indifferent, crystallizable euphorbon. The milky juices which here come under consideration are white in their fresh condition.

Caoutchouc appears in the laticiferous juices in the form of globules, which swell in volatile oils, and are not changed by

[1] Compare among others S. Dietz, " Beitrag zur Kenntniss des Milchsaftes der Pflanzen," Botan. Centralb., 1883, xvi., p. 133.

dilute alkalies and acids, but are dissolved by chloroform and carbon bisulphide.

With the latex-tubes which are formed through solution of the intervening cell-walls are connected the **receptacles for resin and balsam,** which are produced in a lysigenic manner (see pages 170 and 248).

These appear as roundish spaces or cavities, filled with their contents (formerly termed "interior glands"), and are produced in such a manner that those cells which occupy the place of the subsequent receptacle become filled at an early period with the respective secretion ; afterwards the membrane of these cells containing the secretion disappear (Fig. 168, com-

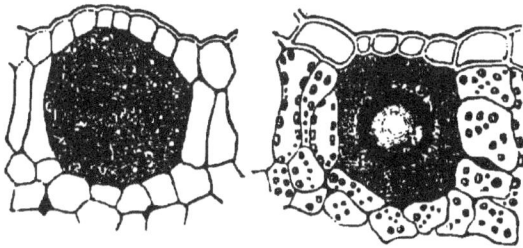

FIG. 168.—Formation of an oil gland of *Dictamnus Fraxinella;* at the left is represented, below the epidermis, a group of small cells which become filled with drops of oil, while the cells are in process of dissolution ; at the right, most of these cells have already become dissolved, and in their place is produced a lysigenic intercellular space containing a secretion (Rauter).

pare also the previous remarks in connection with the cell-membrane). These receptacles are not, like the resin-canals (which will be considered directly), bordered by a circle of secreting cells; the cells inclosing them are not essentially distinguished from the tissue of their surroundings. With these may be classed the oil receptacles of the Rutaceæ[1] (*Ptelea,*

---

[1] Compare Rauter, "Zur Entwickelungsgeschichte einiger Trichomgebilde." Sitzungsber. d. Wiener Akad., 1872, and Von Höhnel, "Anatomische Untersuchungen über einige Secretionsorgane der Pflanzen." Wiener Akademie, November, 1881.

Correa, Ruta, Dictamnus),[1] of the leaves and fruits of *Citrus*, and of *Jaborandi* leaves.

Thus, for example, in the very large oil-spaces in the rind of the fruit of species of *Citrus*, a solution of the cell-walls is distinctly perceptible.[2] This is, perhaps, still more the case in the trunks of *Copaifera*,[3] in which the balsam passages attain an enormous development. These trees contain the *copaiva-balsam* in canals which are as much as an inch in diameter, and which often traverse the entire trunk, so that a single one, after being bored, is capable of yielding balsam by the pound. In the so-called "gummosis," the membranes become converted into gum (compare page 166, Fig. 78).

In the Sterculiaceæ, lysigenic (protogenic, page 263) gum-passages are found.[4]

The lysigenic passages, like the schizogenic, which will be described directly, may be either *dermatogenic*, that is, produced by the participation of epidermal cells (*Citrus, Dictamnus, Amorpha*), or they may be formed under the epidermis, deeply in the interior (interior glands in a more restricted sense).

The secreting space of the lysigenic passages is always completely closed.

The **intercellular receptacles for secretions**, or the so-called *oil-passages* and *balsam-passages*, belong neither to the true cells, nor to the tubes produced through the fusion of cells, nor to the passages of lysigenic origin.

The formation of these receptacles, which are found in the Myrtaceæ (*Eucalyptus*, Figs. 127, 128, *Myrtus, Eugenia, Pimenta*), in the Leguminosæ (*Amorpha, Hymenæa, Trachylobium*), in the Umbelliferæ, Compositæ, and Coniferæ, in *Oxalis, Lysimachia* and *Myrsine*,[5] admits, in the families of the Um-

[1] Here also in the interior of hairs, Sachs, "Lehrbuch," p. 93. Martinet, *loc. cit.* De Bary, *loc. cit.* Fig. 22.

[2] Sachs, "Lehrbuch der Botanik," 1874, p. 92.

[3] Karsten, Botan. Zeitung, 1857, p. 316.

[4] Trécul, Compt. rend., 1862, p. 315.—Ledig, Botan. Centralb., 1881, vi., p. 387.

[5] Von Höhnel, in the investigation cited on page 250.

belliferæ, Compositæ and Coniferæ [1] (which are very rich in such examples) of being traced back to the production, extension, and prolongation of intercellular spaces. Very frequently four cells

 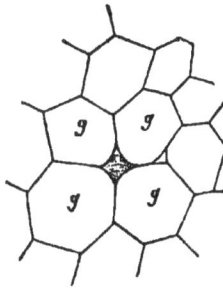 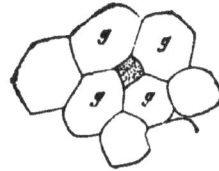

FIG. 169. FIG. 170. FIG. 171.

FIGS. 169–171.—Four boundary cells, *gggg*, in Fig. 170 receding from each other; in Fig. 171 the intercellular passage so produced exerts a pressure upon the boundary cells.

Fig. 169, *gggg*) recede from each other in the region where they meet together.

The boundary cells (*g*), which in the beginning (Fig. 170) still project with their convex walls into the resin-passage,

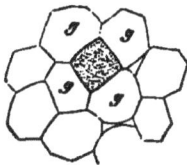 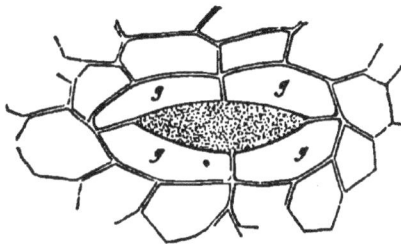

FIG. 172. FIG. 173.

FIGS. 172–173.—Beginning of the depression of the boundary cells, which in Fig. 173 are strongly compressed in a radial direction (Müller).

recede (Fig. 171), and are more and more depressed in a radial direction (Figs. 172, 173). At the same time, in the boundary cells, and frequently also in the farther surroundings of the

[1] Compare also Fig. 106.

passages, a new formation of cells takes place by the division of the older cells (Figs. 174, 175), so that finally the fully devel-

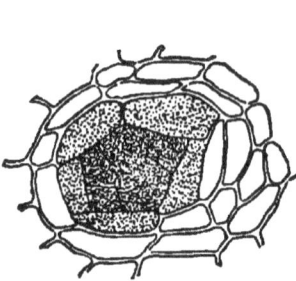
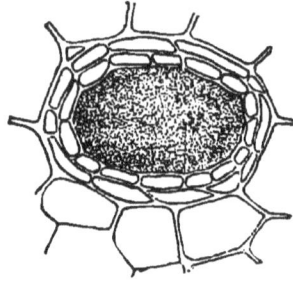

FIG. 174.   FIG. 175.

FIG. 174.—Beginning of the new formation of cells in the boundary cells (Müller).

FIG. 175.—Around the intercellular passage, which has become extended as an oil-space, a layer of tabular daughter-cells (secerning cells) has been formed. Transverse section through a branch of the root of *Inula Helenium* (Müller).

oped resin-passage or balsam-passage is located in a special

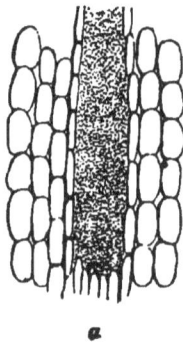

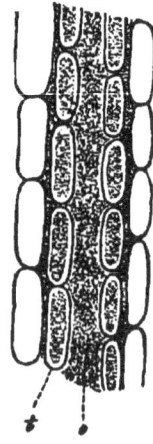

a   *b*   *t*   *o*

FIG. 176.   FIG. 177.

FIG. 176.—Longitudinal section (*a*) and transverse section (*b*) through a balsam-passage (oil-space) of *Rad. Inulæ*.

FIG. 177.—Central portion of an oil-space from *Rhizoma] Arnicæ*, on a longitudinal section; *o*, oil-space; *t*, daughter-cells which have not yet become depressed.

form of tissue (Fig. 176). From these cells surrounding the intercellular balsam-passage (secerning cells, epithelium) the volatile oil, and the resin dissolved therein, penetrate through their walls into the passage.

A canal which has been formed in the manner just described, by the parting of cells from each other, is termed *schizogenic*.[1] In the *larch*, in addition to resin receptacles of this kind, lysigenic passages are also found (page 248).

A ramification of those passages which more closely concern us here is not perceptible on a longitudinal section through them,

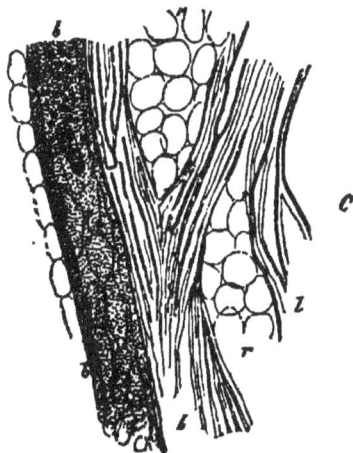

Fig. 178.—Longitudinal section from the bark of *Radix Sumbul* (*Ferula* seu *Euryangium Sumbul*); *r*, medullary rays; *l*, phloëm; *b*, oil-passage (the parenchyma lying beneath it being visible).

so that they do not represent a vascular system, like the latex-tubes of many plants (*e. g.*, Figs. 163–165). The more simple form of the resin-passages is in harmony with the manner of their formation, although this does not exclude the passages from occasionally attaining a considerable length (Fig. 176), as in the *sumbul root* (Fig. 178), in *Rhizoma Imperatoriæ*, or, as already mentioned, in *Copaifera*, and presumably also in *Dipterocarpus*.[2]

[1] Σχίζω I cleave, and γένος production.
[2] Flückiger, " Pharmakognosie," p. 86.

The intercullar resin-receptacles are widely distributed in the Coniferæ; here they are found not only in the bark (Fig. 181), but also in the wood (Fig. 180), in the scales of the cones,[1] and, indeed, the leaves mostly contain two of them (Fig. 179) on the lower surface. It is only in a few Coniferæ (*Taxus*) that they are entirely wanting.

Copal is also produced in schizogenic secretion-receptacles, in the *Trachylobium* as well as in the *Hymenæa*, as Höhnel has shown.

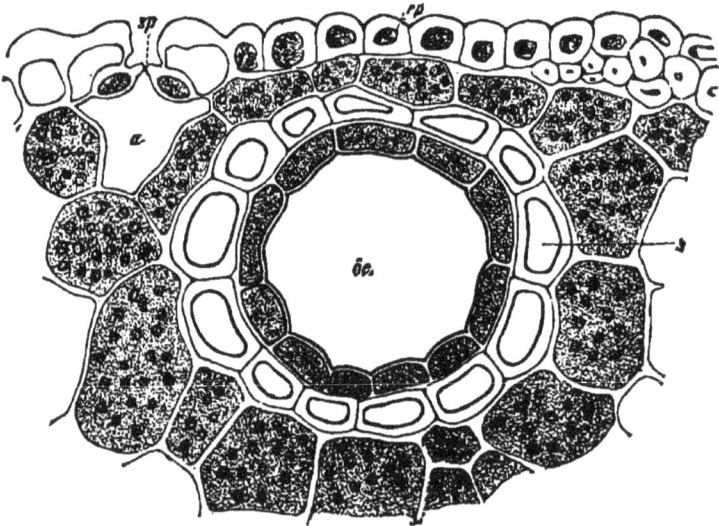

Fig. 179.—Transverse section through the oil-canal of a *coniferous leaf: δc*, oil canal; *s*, nucleus-sheath (consisting of bast-cells) surrounding the same; *sp*, stoma; *a*, respiratory cavity; *ep*, epidermis (Tschirch).

Although a longitudinal section through the root structures of the Compositæ and Umbelliferæ does not reveal either a regular arrangement[2] or a connection of the balsam-passages

[1] Hanausek, " Ueber die Harzgänge in den Zapfenschuppen einiger Coniferen." Jahresbericht der, etc., Handelsschule in Krems, 1880.

[2] With regard to the distribution of the resin-passages, compare Hanausek, " Zur Lage der Harzgänge," Irmischia, ii. (1882), Nos. 3, 4, and N. J. C. Müller, " Untersuchung über die Vertheilung der Harze, äther.

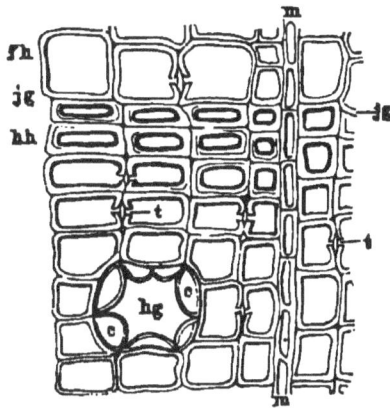

Fig. 180.—Transverse section from the wood of *Pinus silvestris; jg*, annual ring: *fh*, spring wood; *hh*, autumn wood; *hg*, resin-canal; *c*, secreting cells; *t*, pits; *m*, medullary rays.

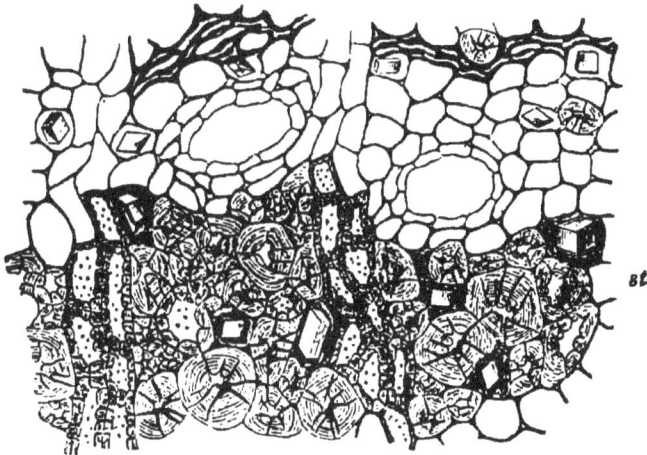

Fig. 181.—Transverse section through the bark of *Pistacia Lentiscus; st*, plate of stone-cells with inclosed crystal-cells, the medullary rays sclerotized.   Resin-passages.

Oele, Gummi und Gummiharze und die Stellung der Secretbehälter im Pflanzenkörper," in Pringsheim's Jahrb. für wissenschaftl. Botanik, v. (1867), p. 380.   H. Mayr, " Entstehung und Vertheilung der Sekretions-organe der Fichte und Lärche," Botan. Centralblatt, xx. (1884), 278.

among themselves; yet a conformity to some law in the position of these receptacles is sometimes apparent upon a transverse ‸section. The root-stock of *Arnica*, for example, shows a large balsam-passage (Fig. 182) before each fibro-vascular bundle. A similar arrangement may be seen to exist in

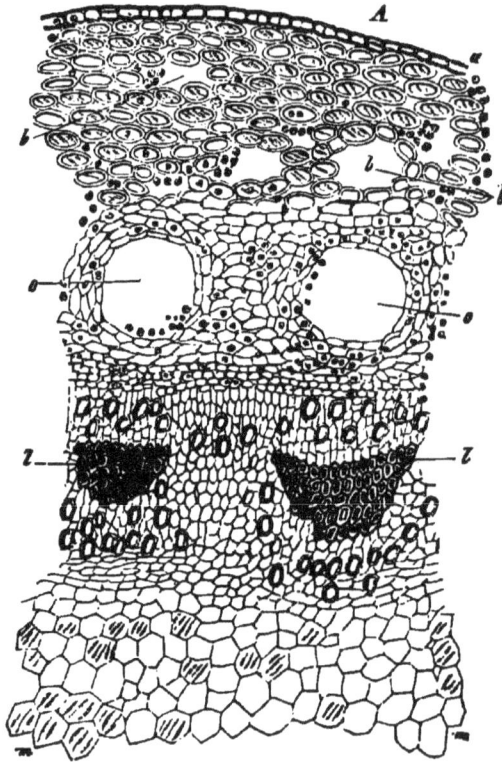

Fig. 182.—Portion of a transverse section from the underground stem (rhizome) of *Arnica montana*. Before each xylem-ray (wood-bundle) *l*, there is a very large oil space; *o, b*, cavities formed by the laceration of the fundamental tissue; *a*, epidermis of the root (epiblema); *m*, medulla. The cells of the epidermis are spirally striped; in the surroundings of the oil-spaces are drops of oil which have escaped. Compare also Berg's " Atlas," Plates 8, 9, 10.

Radix Angelicæ and *R. Levistici*, in *Rhizoma Imperatoriæ* and others, although, in the course of development, which is not

17

always perfectly uniform, they often become disturbed in consequence of the laceration of the individual tissues.

Hand in hand with the organic transformations just described, there occur, in the surroundings of the passages, certain chemical processes, to which the resins, volatile oils and varieties of mucilage owe their origin and also the special form which enables them to pass into the intercellular spaces. The resins, namely, are either dissolved involatile oils, as " balsams " or "turpentines," or they are emulsified by mucilage (gum). It is only in this form that they are capable of passing through the cellwalls into the passages formed for their reception. Although the manner of formation of the cells and tissues [1] which have here been considered may appear quite clear, yet the chemical side of these processes has so far not been elucidated. In many cases, resin and volatile oil appear to be produced from amylum. If this may be accepted with some degree of probability, the equally just supposition is forcibly presented that under certain conditions cellulose, which agrees with amylum in its composition, is also capable of undergoing the same transformation. As a matter of fact, this is also the case with the lysigenic canals, which have previously been considered (compare page 248).

According to Frank's investigations,[2] it would appear as if the oil-tubes or stripes, vittæ,[3] which are so characteristic for many of the umbelliferous fruits, first became forced asunder through the volatile oil which makes its appearance in them, while the balsam-passages of umbelliferous roots present the development illustrated by Figs. 169 to 175. But in many of these fruits the oil-tubes also show remnants of transverse walls (Fig. 183) which presumably indicate a solution of original boundary cells. The effloresced appearance of the tissues which surround the oil-tubes in Fructus Carvi, Fructus Fœniculi,

[1] Especially described by Müller, loc. cit., p. 387.—Thomas, Ibid., v., p. 48.—See also Frank, " Beiträge zur Pflanzenphysiologie," Leipzig, 1868, pp. 120, 123.
[2] " Beitr. z. Pflanzenphysiologie," p. 128.
[3] Fig. 94 o.—Berg's " Atlas," Plates xli., xlii., xliii.

etc., also supports this view.   It has, however, recently been
shown that the oil-tubes are produced in the same manner as
the resin-receptacles of roots,[1] and are thus of schizogenic
origin.

As an exception, in *Fructus Conii* oil-tubes[2] are not found
at the period of ripening (as has previously been already mentioned), but rather a connected layer of cells (Fig. 184), in
which is located the volatile oil and coniine.   When observed

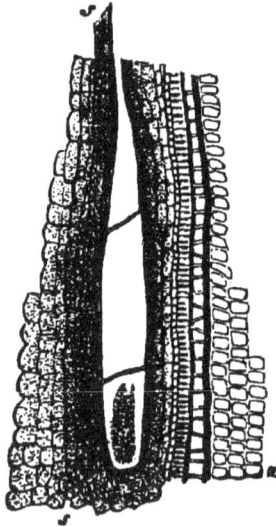

Fig. 184.—Longitudinal section through an oil-passage from *Fructus Fœniculi*, with
transverse walls; *s*, dark-brown effloresing cork-like tissue; *a*, albumen of the seed.

on a longitudinal section (Fig. 185), this layer represents the
cells superposed in a compartment-like form; if their transverse

[1] Lange, "Über die Entwickelung der Oelbehälter bei den Umbelli-
feren," Königsberg, 1884.   Dissertation.—Compare also E. Bartsch,
"Beiträge zur Anatomie und Entwickelungsgeschichte der Umbelli-
ferenfrüchte." Inaugural Dissertation, Breslau, 1882.

[2] In young ovaries they are present as an initial formation, but do not
become developed.

walls were to disappear, the figure of an oil-passage produced
by resorption would be obtained. Such a solution of the transverse walls does not occur, however, in *Fructus Conii.*

The spaces which have just been considered contain either
volatile oil (Myrtaceæ), which in drugs is often already more or
less resinified, or a mixture of volatile oil and resin, or finally
resin itself.

Resin, which is free from volatile oil, is also met with in
Lignum Guaiaci and in *Lig. Quassiæ* in the form of brittle,

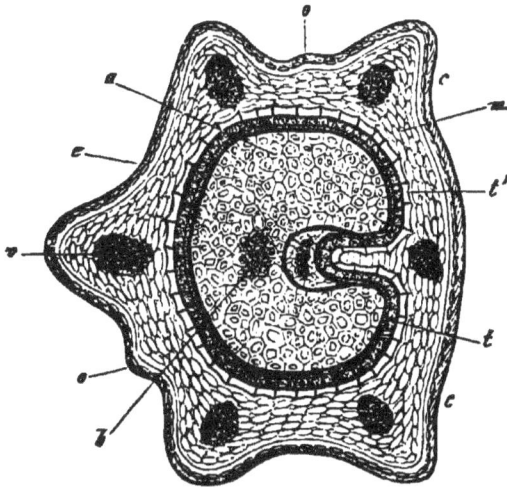

Fig. 184.—Transverse section through *Fructus Conii; a,* albumen of the seed; *b,*
embryo; *cc,* commissural surface; *e,* epidermis; *m,* tissue of the fruit casing; *t',* innermost layer of the latter; *t.* layer of cells containing volatile oil and coniine; *o,* vittæ; *v,*
ribs (costæ), traversed by fibro-vascular bundles.

shapeless masses. In these two plants the resin first makes its
appearance at a later period (at an advanced age) in ordinary
wood-cells and vessels of the heart-wood; it thus shows a deportment similar to that of physiological gum (see page 166), with
which it also appears to be chemically related (see also page
264).

By a microscopical examination, the resin and oil, especially
when they occur mixed with each other as a balsam, escape from

these depositories in the form of small, strongly refracting drops, which are more frequently yellowish or brownish than colorless, and are clearly miscible with alcohol, or at least with absolute alcohol, ether or benzol, and with fixed or volatile oils.

A mixture of gum, resin, and volatile oils (gum-resin), as in *asafœtida, galbanum*, and *ammoniacum*, or even gum or mucilage alone (Cycadeæ),[1] is also found in schizogenic secretion-receptacles.

However, should any of the material occurring in cells or in

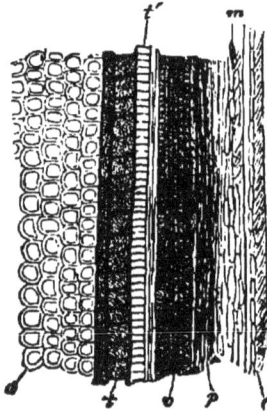

FIG. 185.—Longitudinal section through the coniine layer, *t*, of Fig. 184. The letters have the same signification as in Fig. 184.

intercellular spaces be found to be indifferent towards the above solvents, it does not necessarily follow that resin is absent, since the older lumps of resin dissolve only with difficulty.

The resins are colored red by tincture of alkanet. The reagent of Unverdorben and Franchimont (an aqueous solution of acetate of copper) colors the drops of resin, after several days' maceration in the liquid, emerald-green.

The resins are not, however, confined to these receptacles, as

[1] G. Kraus, in Pringsheim's Jahrb. f. wissensch. Botan., iv., p. 305, Plates xxi. and xxiii.

is already evident from the description of the latter. In young cells, where the resin is first formed only in small amount, it is destitute of any special color; the same is the case where the resin forms semi-liquid granules, especially in those plants where it occurs emulsified as a constituent of laticiferous juices, for example, in *Tuber Jalapæ*. This extremely fine division and liquefaction of the resins is promoted in the laticiferous juices of the Umbelliferæ by the volatile oil which they contain. Under such conditions the resin may be recognized by its tendency to absorb coloring matters. Iodine solution, or preferably aniline colors dissolved in water, carmine, etc., when carefully added in corresponding amount, are very useful for this purpose. To be sure, this does not always prove that the colored bodies are resins, but the coloration, nevertheless, affords good points of discrimination.

The schizogenic secretion-receptacles are generally closed. There are, however, cases where the secreting space is open, and communicates with intercellular spaces of the parenchyma, which likewise sometimes contain a secretion (*Oxalis floribunda, Peganum Harmala, Lysimachia Ephemerum*).

With regard to their morphology and the nature of their development, the following varieties of secretion-receptacles should, therefore, be maintained distinct:

I. *True Cells.*

(*a*) Isolated in the interior of the tissues, and on all sides in connection with the other elements of the tissue.

    1. Containing oil : *Macis, Laurus, Cortex Angosturæ,* roots of the Zingiberaceæ, *Acorus Calamus, Cinnamon bark.*

    2. Containing mucilage : *Cinnamon-bark, Elm-bark, Althæa root* (filling entire tissues in *Chondrus*).

    3. Containing laticiferous juice, simple : *galls,* from species of *Rhus,* from Eastern Asia : Euphorbiaceæ.

    4. Containing crystals, especially in monocotyledons.

    5. Containing tannin.

(*b*) United in a band of cells. *Fructus Conii, Tuber Jalapæ, Dichopsis Gutta.*

(*c*) As glandular heads of hairs.
1. Upon the epidermis, glandular hairs of the Labiatæ. *Kamala, Glandulæ Lupuli* (colleters).
2. Projecting into intercellular spaces (*Aspidium Filix mas*).

II. *Secretion-receptacles produced through the fusion of cells:*
(*a*) Of rows of cells : latex-tubes (Papaveraceæ, Cichoriaceæ, Campanulaceæ).
(*b*) Of homogeneous aggregations of cells : lysigenic oil- and balsam-passages (Rutaceæ, including the Aurantieæ).
(*c*) Of entire, and also dissimilar portions of tissue : gum-glands.

III. *True intercellular spaces :*
(*a*) Containing oil or resin, schizogenic balsam-passages (in the Coniferæ, Umbelliferæ,¹ Myrtaceæ, Leguminosæ, Hypericineæ).
(*b*) Containing gum-resin (the roots of some Umbelliferæ).
(*c*) Containing mucilage, schizogenic mucilage-passages (Cycadeæ).
(*d*) Containing laticiferous juice (*Alisma Plantago*).

The secretion receptacles may be *protogenic,* that is, they may be formed already in quite young tissues, or they may be *hysterogenic,* that is produced at a later period in old and completely developed tissue.

In transmitted light, many leaves appear finely punctate in consequence of the presence of resin cells, resin cavities, and crystal cells. Bokorny has utilized these "pellucid points" diagnostically.² The pellucid points are caused by resin cells in the Lauraceæ, Piperaceæ, Meliaceæ, Sapindaceæ, Canellaceæ, Anonaceæ and others, by resin-cavities in the Myrsineæ, Myrtaceæ, Rutaceæ, and Hypericineæ.

¹ The size of the resin-canals possesses diagnostic value in distinguishing *Rad. Levistici* and *Angelicæ.* In the former they have the same diameter as the vessels ; in *Angelica,* on the contrary, they are considerably wider.

² " Die durchsichtigen Puncte der Blätter in anatomischer und systematischer Beziehung." Flora, 1882.

PATHOLOGICAL FORMATIONS.

The morphological and anatomico-physiological relations of the organs formed in the normal vital processes of plants having been considered so far as they relate to our purpose, some other phenomena may still be mentioned which are produced by disturbances of the normal processes.

When a part of a plant is wounded by the human hand or by an animal, it is capable of repairing the injury. The most usual form of protection is the formation of cork (protective cork) on the wounded place. Figure 109 *a* shows, for example, how on a fruit of the *Vanilla*, which has been wounded by an insect, cork has been produced around the wounded place, whereby access of air is excluded from the inner tissues. Cork is the ordinary form of protection for delicate organs, but naturally, is only met with in such places where cells occur which are still delicate and readily suberized. If, on the contrary, the stem of a dicotyledonous woody plant is wounded down to the wood, the plant selects another means of protection, since cells here become exposed which can no longer become suberized. Accordingly, in all the cells of the wood which border on the wounded place, gum (wound-gum) is produced as an exudation of the thick membranes, which gradually fills the cell cavity and thus directly closes the wound. The masses of gum and resin which occur in the dead heart-wood of the officinal woods (page 260) are probably such protective gum. During this process, in those portions of the bark which are still capable of development, the plant endeavors to close the wound from both sides. There are thus produced, by very active growth in the contiguous cambial zone, broad, inflated cushions, or *excrescences,* which gradually constrict the wounded place more and more, and, indeed, may finally entirely close it.

Such enlargements of the tissue as are produced by outward influences are termed *hypertrophies.* They occur not only in the form of excrescences, but also as variously shaped malformations, which should not be classed with the excrescences without further distinction. While, namely, the latter are to be regarded as a reaction produced in consequence of a single wound, the other anomalous formations, quite generally designated by the name of *galls* or *cecidiæ,* are of service to the insect or parasitic plant which produces the wound. Hence there may be distinguished according to the respective organism, *fungus-galls* (mycocecidiæ)—for of all plants only the parasitic fungi produce such formations—and *insect-galls* (zoocecidiæ.) Whether, however, a plant or an insect is the cause of these formations, the symptoms are always the same. By a vigorous, and often greatly increased formation and division of cells, the affected part of the plant, which always consists of cells which are still capable of development, is very considerably enlarged, and manifold and often very strange malformations are produced. An active conveyance of sap affords abundant nourishment, which is often so great that not only are the requirements fulfilled for the new formation of cellular tissue, but numerous surplus products (for instance, starch) also accumulate in the cells. Moreover, peculiar bodies (such as tannic matters) are also frequently formed in the cells of the gall, which are either wanting in the other tissues of the plant, or are contained in them in very much smaller amount. Upon such an activity of development, which is increased and qualitatively changed through irritation, essentially depends the formation of galls.

In the fungus-galls, the hyphæ of the fungus, nourished by the richly supplied cells, penetrate the intumescence. In the case of the insect-galls, the interior, which is mostly hollow, serves as an abode for the insect; it there passes through its entire course of development, from the egg to the perfect insect (*Cynips* galls of the *Oak*), and even through several generations. Since the galls seriously injure the plant only when they are produced in quantitatively very considerable amounts, a case of *symbiosis*[1] (cohabitation) is thus presented here, of beings belonging to two different series, though, since the one lives chiefly at the expense of the other, this borders closely on parasitism. These formations are very remarkable from the fact that a plant does service for an insect and constructs for the latter its dwelling place.

Since insects of different classes (Hemiptera, Diptera, Hymenoptera) participate in the production of insect-galls—and only these interest us here— it is impossible to make any statements of general application regarding these gall formations, which occur upon plants of all the families of phanerogams.[2] Their shape, like the nature of the irritation which produces them, is extremely variable.

Only those galls which are rich in tannin are of technical importance, and these are also our best sources of tannin. The *oaks*, especially, furnish many valuable galls.[3]

The galls of Asia Minor (or galls in a general sense) are produced by the puncture of the ovipositor of the female insect of *Cynips gallæ tinctoriæ* Olivier (a Hymenoptera) in the young shoots of *Quercus lusitanica* Lamarck. The female insect, developed in the hypertrophic tissue from the egg deposited therein, afterwards bores for itself a passage and escapes from the gall, by which it was sheltered during one of its phases of life.

The so-called *Chinese* and *Japanese galls,* on the contrary, are produced by the female *Aphis sinensis* (a Hemiptera) in the younger shoots and leaf-stalks of *Rhus semialata* Murray. In these galls, which are mostly very large, the numerously introduced eggs become developed as plant-lice, pass through succes-

---

[1] Σύν with, and βίειν to live.

[2] Compare especially, Frank, "Handbuch der Pflanzenkrankheiten."

[3] The galls which are employed technically and pharmaceutically have been very excellently described by Hartwich (Arch. d. Pharm., 21, 1883, p. 820); compare also Wiesner, "Rohstoffe des Pflanzenreiches," Vienna, 1873, pp. 846, with numerous illustrations.

sive generations, and finally escape. In the commercial galls, the small insects are still frequently found which have been killed by scalding or have otherwise been destroyed.

MICRO-CHEMICAL REAGENTS.

In the course of the preceding representation the chemical detection of one or another substance has already been alluded to, and, indeed, the treatment of microscopic sections with suitable reagents affords many valuable disclosures. As in all other cases, definite answers are obtainable, when systematic and accurately formulated questions are propounded. For this purpose, chemical reagents [1] are of service, among which the following may be designated as especially important:

1. **Chromic Acid** (free from sulphuric acid) dissolved in 100 parts of water. This is adapted in general for the purpose of loosening composite, thickened cell-walls and constituent bodies, whereby the finer relations of structure are very often made evident, since chromic acid is also capable of clearing up darkly colored cell-walls, and, on the other hand separates the layers and finer membranes, thus bringing them more clearly to view. By means of this acid, starch granules are completely separated into laminæ, the strata of the *Cinchona* bast-fibres (page 155, Fig. 73) separated from each other and lignified membranes dissolved, while suberized membranes are rendered more clear.

When employed in a more concentrated form, or allowed to act for a longer time, chromic acid destroys the cell-walls. Its application, therefore, requires continual observation of the sections which are treated therewith, in order to completely survey the result of the phenomena.

[1] Compare Poulsen's "Botanische Microchemie," Cassel, 1881; and Behrens' "Hilfsbuch zur Ausführung microscopischer Untersuchungen," Braunschweig, 1883. The American editions of these two works are noticed on page 49; in both of them are contained numerous references to the literature of the subject.

2. **Hydrochloric Acid** of the specific gravity 1.110 acts much less energetically upon the cell-walls; yet, without causing these to swell in a disturbing manner, it seizes upon many of the constituent substances and thereby permits the structure of the tissue to be more clearly recognized (a stronger acid effects tumefaction). Calcium oxalate (page 129) is readily dissolved by hydrochloric acid.

3. **Sulphuric Acid.**

The dilute acid (specific gravity 1.110) causes starch and the membranes to swell. Cellulose is converted by it into amyloid (page 159).

Concentrated sulphuric acid (specific gravity 1.836) dissolve the membranes and their contents; only the cuticle, cork, the nucleus sheaths (page 202), intercellular substance, and the oil drops contained in the cell resist its action.

In the phloroglucin reaction (page 161), sulphuric acid can be applied in place of hydrochloric acid. When mixed with indol [1] the former is a good reagent for lignified membranes.

4. **Nitric Acid** of the specific gravity 1.180, either alone or after the addition of ammonia, colors protein substances, as also the middle lamella, yellow. Höhnel employs it for the detection of suberin (cerin reaction).

Since nitric acid dissolves starch as well as sulphuric and hydrochloric acids, it can likewise be employed for clearing up tissues which are rich in starch.

Nitric acid alone, or, still better, a mixture of nitric acid and potassium chlorate (Schultze's maceration), is the best agent for isolating the individual element of tissue. The boiling acid, to which small crystals of potassium chlorate are gradually added, dissolves the middle lamella. This method of procedure is particularly applicable to the examination of vegetable powders (*Cinnamon, Cinchona, etc.*); dark membranes at the same time become bleached thereby.

When treated in this manner, boiled scraps of drugs are re-

[1] A colorless crystalline principle of the composition C_8H_7N, obtained by the action of reducing agents on the blue coloring principle of indigo. F. B. P.

solved by the pressure of the glass cover upon the slide into the separate elements (Fig. 186), which may then be conveniently further examined. It is, however, to be considered that the reagent produces a swelling of the membranes, as also a solution of the bodies contained therein. It is likewise to be observed that Schultze's maceration dissolves the lignin from the lignified membranes, so that the latter then show the cellulose reaction.

Before the objects that have been thus treated are brought

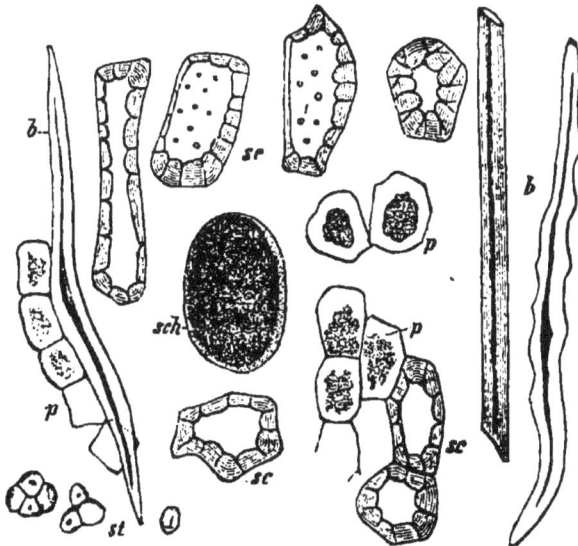

Fig. 186.—Isolated elements of *cinnamon bark* obtained by maceration with Schultze's liquid. *b*, bast-cells; *sc*, stone-cells (sclereids); *p*, parenchyma; *sch*, mucilage cells; *st*, starch granules of the *cinnamon* (Tschirch).

under the microscope, it is necessary to thoroughly wash out the reagent.

5. **Acetic Acid** of the specific gravity 1.040 often clears up in a remarkable manner such sections as have previously been treated with alkalies. Since calcium oxalate is insoluble in acetic acid, the latter may also be employed as a confirmative test when the recognition of this salt is in question (compare page 133). This acid is likewise of service in the examination

of protoplasm, since it causes the nucleus of the cell to appear sharply defined.

6. Neutral **Acetate of Copper** dissolved in twenty parts of water. If tissues containing resin are allowed to remain for some days in this solution, the metal combines with the resin to form green clots.

7. **Tannic Acid** is always dissolved freshly just before use in twenty parts of water, and employed in the search for alkaloids.

By moistening thin sections with water, to which a slight trace of acetic acid has been added, a concentrated liquid extract of the substance is prepared, which is tested by the gradual addition of a few drops of the solution of tannic acid upon the slide. If a turbidity is produced, it may be due to the presence of alkaloids, but may also be caused by the so-called bitter principles, or by albumen.

8. **Solution of Soda** of the specific gravity 1.160, containing 15 per cent of sodium hydroxide. Instead of this solution, an equally strong alcoholic solution may serve for many purposes. The corresponding solutions of potassium hydroxide have the same action.

The caustic alkali causes the cell-walls to swell, and dissolves many of the constituent substances, especially coloring matters, whereby the sections are rendered very much clearer; from lignified membranes the lignin is extracted by means of warm solutions of caustic potassa or soda. In many cases the treatment of the tissue with alkali permits of the clearer recognition of the relations of the strata. The protein crystalloids reveal their organic structure by swelling, whereby their plain surfaces become for the most part rounded and the angles changed. Many yellow coloring substances (chrysophan in *Rhubarb*, frangulin in *Cortex Frangulæ*, chrysarobin) become red by alkalies, a reaction for which lime-water is well adapted.

9. **Sodium Hydroxide** (solid caustic soda) or potassium hydroxide may be conveniently kept and employed in the form of powder, when the amount to be applied does not require to be more accurately measured.

10. **Ammonia Water,** of the specific gravity 0.960, is often

preferable to potassa and soda, since the two latter frequently produce an altogether too energetic swelling, which detracts from the clearness of the outlines. The jelly which is formed by the action of the alkalies upon starch also interferes very much. By the application of ammonia neither of these objectionable results are produced, while its power of dissolving coloring substances is not less in extent. Starch suffers no change by the action of ammonia.

Ammonia water, still further diluted, is adapted for softening dried plants and many drugs which it is desired to examine more thoroughly. Sections which are treated with nitric acid, and afterward washed and moistened with ammonia, admit of the sharp recognition of protein substances and of the middle lamella by means of their yellow color (xantho-protein reaction).

11. **Alkaline Solution of Tartrate of Copper.** The solution of the tartrate of copper and sodium in caustic alkali, the so-called " Fehling's solution," is not convenient for microchemical purposes. In place of it, the following method of procedure may be recommended: A solution of 3 parts of sulphate of copper (blue vitriol), free from iron, in 30 parts of hot water is mixed with a solution of 7 parts of Rochelle salt (potassio-sodium tartrate) in 20 parts of hot water, the resulting precipitate collected and dried. When used, a little of this precipitate is brought upon the object-glass, a small fragment of caustic soda added, and thereupon a few drops of water until a clear solution is produced, or this may also be effected by the use of the least possibly quantity of the solution of caustic alkali (No. 8). The section is then moistened therewith. This alkaline solution of the tartrate of copper is useful in testing for sugar, since uncrystallizable, so-called fruit sugar (page 142) immediately separates therefrom reddish-yellow, hydrated cuprous oxide. This also occurs very soon by the aid of a gentle heat when grape sugar is present, but not even by boiling in the case of cane sugar (or mannite). Dextrin is also capable of reducing the tartrate of copper with the aid of heat.

The varieties of gum and mucilage effect no reduction in the alkaline solution of tartrate of copper.

Sachs proceeds in the following manner in the reaction for grape sugar. He places the (thick) longitudinal sections for some minutes in a solution of sulphate of copper (1 part of sulphate of copper and 4 parts of water), then washes them with water, and brings them into a boiling solution of caustic potassa (one part of the solution No. 8 and two parts of water). Cells containing grape sugar then appear filled with a reddish-yellow, granular precipitate (Cu_2O). In this reaction it is necessary to accurately proportion the time that the section remains in the liquid, the amount of washing, thickness of the section, etc., for which some experience is required.

The alkaline tartrate of copper imparts to the albuminous substances deposited in the parenchyma a violet color, in consequence of the formation of compounds of copper with the protein substances, as was made known in 1872 by Ritthausen.

12. **Ammoniacal Solution of Oxide of Copper** is obtained by shaking copper turnings with ammonia water of the specific gravity 0.960, with the addition of very little ammonium chloride. The ammonical oxide of copper, when prepared in another manner, has a different action in some cases. This liquid is the only solvent for cellulose. It is to be observed, however, that its action upon the cell-walls is very different, according to their thickness and purity, and that many, as for instance the hyphæ of fungi and cork, are not attacked by it at all, or at least not without previous boiling with caustic alkali or with potassium chlorate and hydrochloric acid. The action of the ammonical oxide of copper does not occur immediately.

The ammoniacal oxide of copper is only fit for use when it dissolves cotton in the course of a few hours. It is expedient to protect it from the action of light, and not to keep it for a very long time.

13. **Glycerin** of the specific gravity 1.225 is of very general use as a clearing agent; with a higher degree of concentration its power of abstracting water also comes into consideration. In the examination of such constituent substances as would dissolve quickly in water (aleurone, tannic acid), concentrated glycerin is very useful, since its solvent power is only gradually

18

exerted. Thus under glycerin the gradual swelling of membranes affording mucilage and the disintegration of tissues containing oil may also be conveniently observed, and by the addition of water accelerated at will.

14. **Absolute Alcohol** is of service, for example, in rendering
mucilage visible, which would be washed away by water or would
form a clear mixture with glycerin. The volatile oils and
resins are dissolved by alcohol.

Fats and wax are but slightly soluble in cold alcohol, but, for
the most part, can be brought into solution by boiling.

Protoplasm is killed and hardened by alcohol. Since the latter has a dehydrating action, its application, either alone or with
the addition of glycerin, effects a separation of the protoplasm
from the membrane.

The tissues can be freed of air by means of alcohol, since the
latter more readily penetrates into the intercellular spaces, and
is also capable of absorbing more air than water.

By placing the respective organs in alcohol, inulin (page 124)
as also hesperidin (page 143) are obtained in sphæro-crystals.
Indeed, by means of alcohol, even asparagin and sugar may be
made to crystallize in organs which are particularly rich therein.

15. **Alcohol of 85 per cent** by weight, having a specific
gravity of about 0.830, the ordinary Spirit of Wine, accomplishes
in most cases the some purpose as absolute alcohol.

16. **Alcohol of 60 per cent**, besides dissolving the resins,
also dissolves the different sugars in considerable amount.

17. **Ether** is applied for the removal of solid and liquid fats,
whereby resins and volatile oils are dissolved at the same time.

18. **Benzol** (C_6H_6) serves the same purpose as ether, but,
since it boils at 80° C., it admits of gentle warming (with care!),
which is often very useful. The same may be said of chloroform.

19. **Chloroform.** Resins, fats, wax, and volatile oils dissolve
in ether, benzol, and chloroform. These liquids, as a rule, are not
allowed to flow upon the preparation (under the glass cover of
the slide). It is preferable to place the preparation in a small
watch-glass filled with the reagent.

Hardened masses of resin, such as are frequently found in older drugs, often resist for a long time the action of the solvent, a fact which must be considered in order to avoid incorrect conclusions. An alcoholic solution of caustic soda often has a better action than alcohol and other solvents.

20. **Paraffin** of a low boiling point (55 to 75° C.), the so-called petroleum ether or petroleum benzin. This liquid serves for similar purposes as ether, benzol, and chloroform; it has, however, a much less energetic solvent action upon resins.

21. **Fatty Oil** (Almond Oil is the best) is employed with advantage as a mounting medium, when, beside the fatty oil occurring in the cells, it is desired to examine the other constituents which would become decomposed or dissolved by glycerin or water (aleurone). Sections of seeds rich in oil, when placed in the fatty oil, appear very much clearer, since the oil is taken up by the mounting liquid.

22. **Liquid high-boiling Paraffin** (*Paraffinum liquidum*[1] of the Pharmacopœa Germanica), serves in most cases the same purpose as the fatty oil and is cleaner in its application.

23. **Iodine** in the form of powder, when strewn upon moistened sections, often produces purer colorations than iodine solutions; the excess of iodine is easily washed away with water. Powdered iodine readily cakes together. It is therefore more convenient to use it in form of a trituration with siliceous earth or pumice stone, and to preserve it in this form, since the latter substances are not objectionable in many of the reactions to be made with iodine.

With regard to the application of iodine and its solutions for the recognition of starch, cellulose, and protein substances, compare the respective sections preceding.

24. **Iodine Water.** One part of iodine shaken with 4,000 parts of water is used as a reagent for starch and those forms of cellulose which show a similar reaction.

25. **Iodine Solution** (Iodine with Potassium Iodide) is a

---

[1] A clear, oil-like liquid obtained from petroleum, free from colored, fluorescent, and odorous substances. F. B. P.

solution of 3 parts of iodine and 8 parts of potassium iodide in 1,200 parts of water. After some time, a little hydriodic acid is formed in this solution; the deportment of the solution is then (for example, with starch; see page 122) not precisely the same as when the freshly prepared solution is used.

26. **Iodine Tincture.** A solution of 1 part of iodine in 10 parts of alcohol of the specific gravity 0.830.

27. **Iodine with Glycerin.** A mixture of 1 part of iodine solution (No. 25) with 10 parts of glycerin of the specific gravity 1.230.

28. **Iodine with Chloride of Zinc.** In 100 parts of a solution of chloride of zinc of the specific gravity 1.800 are dissolved 6 parts of potassium iodide and as much iodine (about 1 part) as the liquid is capable of taking up.

Pure cellulose—though not that of the fungi—assumes with the chloride of zinc and iodine a violet color (chloride of zinc causes the formation of amyloid). Cells containing tannin assume with the chloride of zinc and iodine a reddish color.

29. **Potassio-mercuric Iodide** (a solution of mercuric iodide in potassium iodide) is prepared by dissolving 1.35 parts of mercuric chloride (corrosive sublimate) and 5 parts of potassium iodide in 100 parts of water. Nearly all the alkaloids are precipitated by this reagent from their solutions, even when highly diluted, so that it affords indications of the presence of such substances. The precipitated compounds are mostly amorphous, and only a few assume a crystalline form after some hours.

30. **Ferrous Sulphate** (Green Vitriol), prepared in the form of a fine powder, by precipitating it from its solution in water by means of alcohol, and quickly drying it by exposure to the air. When used, 1 part is freshly dissolved in 20 parts of water. Many substances of the class of tannins are colored by this salt, but usually of a different tint than by ferric chloride. The addition of lime-water to sections which have been impregnated with a solution of ferrous sulphate and subsequently rinsed with water, often produces new colorations.

31. **Ferric Chloride.** The officinal solution [Pharm. Germ.]

of the specific gravity 1.405 is diluted with 10 times its weight of water, and the sections allowed to macerate for some time in the liquid. If alcohol be employed for diluting the solution, somewhat different reactions are usually obtained, and, upon the subsequent addition of lime-water, still further changes of color appear. The dilute solution of ferric chloride is decomposed by long keeping (as a result of dissociation). Only the officinal solution should, therefore, be kept ready prepared.

The dilute solution of ferric chloride serves chiefly for the recognition of tannic matters, which are thereby colored either green or blue. It is expedient to discriminate between these two classes of colorations, though this is frequently difficult, owing to the appearance of transition colors in consequence of several tannic matters being usually present at the same time. This assumption of the simultaneous presence of several different tannic matters is supported by the observation that the coloration first produced by very small quantities of solution of ferrous sulphate in cells containing tannin is often changed by the further addition of ferric chloride. The behavior of pyrocatechin, quercitrin, and rutin to iron salts may here also be called to mind.

Instead of ferric chloride, ferric sulphate or ferric acetate may also be employed.

32. **Mercurous Nitrate,** known also by the name of "Millon's reagent." One part of mercury is dissolved, without heat, in 1 part of fuming nitric acid, and the solution diluted with 2 parts of water. This liquid imparts a red color to protein substances, though only when the latter are present in considerable amount. The striping of the membranes is rendered clearer by Millon's reagent. On account of its strongly acid reaction, care must be taken not to have it come in contact with the microscope.

33. **Aniline Sulphate,** in aqueous, or better, alcoholic solution, colors all lignified membranes yellow, especially after the addition of sulphuric or hydrochloric acid.

34. **Phloroglucin** is a still more delicate reagent for lignification. The sections are thoroughly moistened with hydro-

chloric acid, and a freshly prepared solution of phloroglucin in 100 parts of water dropped upon them, whereupon lignified membranes become red. Occasionally, this coloration appears without the addition of phloroglucin, for the reason that this principle itself occurs in some barks.

COLORING AGENTS.

35. Aniline Colors. Fuchsine, methyl-violet, methyl-green, Hanstein's aniline-violet (equal parts of methyl-violet and fuchsine), vesuvine, as also aniline-blue and aniline-brown, receive manifold applications, particularly in bacteriological investigations, since these organisms are capable of strongly absorbing the aniline colors. But in histological investigations the above-named colors are also employed, usually dissolved in 100 parts of water.

36. Eosin in aqueous solution colors dead protoplasm intensely red, and is therefore especially applicable, for example, in the examination of sieve-tubes.

37. Carmine Solution. The best carmine is dissolved in ammonia-water, the clearly decanted liquid evaporated to dryness, and the residue (preferably only as required) dissolved in 100 parts of hot water. This reagent is abundantly absorbed by many substances, for example by albumen and resins, also by the delicate cuticle of cells, so that, by an unequal coloration of the walls and constituent substances, many relations may be rendered clearer.

38. Hæmatoxylin (3.5 parts in 100 parts of water) in combination with alum is an admirable coloring agent for cell-nuclei.

MOUNTING MEDIA.

If it is desired to keep a preparation, it must be preserved in a medium which does not evaporate, and in which the structure of the preparation may be clearly retained. The most convenient to use for this purpose are certain preserving liquids, especially glycerin (specific gravity 1.250) or calcium chloride (one part of

the salt in three parts of water). These two liquids, particularly
the first-named, leave most preparations unchanged, even after
many years. Starch, however, is dissolved by calcium chloride,
even when it is neutral.

For firmer objects, sections of more compact drugs, for
Lycopodium, and also for diatoms (provided they will bear some
warming) *Canada balsam* may be employed as the mounting
medium. The preparations must, however, have previously
been repeatedly washed with alcohol. The section is then placed
in Canada balsam which has been liquefied with a little warm
chloroform, and finally in the slightly warmed balsam itself.

A solution of gelatin in glycerin is also adapted for delicate, as
well as for coarse preparations. This is obtained by gently
warming one part of colorless gelatin with six parts of water
and seven parts of glycerin. When used, the mixture is liquefied by warming.

When Canada balsam or glycerin-gelatin are used, it is not
absolutely necessary to specially cement the cover-glass, since
the solidifying mounting medium holds the cover-glass firmly;
but if glycerin or calcium chloride solutions are used as mounting media, it is necessary to cement the cover-glass. As varnish,
either the ordinary black asphalt varnish or the yellow "prepared gold-size" are employed.

Synopsis of Some of the Previously Mentioned Micro-Chemical Reagents.

REAGENT.	CELLULOSE.	LIGNIFIED MEMBRANE.	SUBERIZED MEMBRANE, CUTICLE.	STARCH.	PROTEIN SUBSTANCES.	TANNIC MATTERS.	SUGAR.
Iodine with Potassium Iodide ("Iodine Solution").	colorless	yellow	brown	blue	orange-yellow.	....	....
Iodine with Zinc Chloride.	violet	yellow	brown	blue (swelling).	yellow	reddish	....
Iodine and Sulphuric Acid.	dissolved with a blue coloration.	slowly dissolved, without a blue coloration.	completely insoluble.	dissolved	destroyed with a brown color	....	....
Ammoniacal Oxide of Copper.	dissolved	undissolved.	undissolved.	....	....	....	....
Alkaline Tartrate of Copper.	dull blue	....	....	swelling	violet	red-brown	reddish-yellow precipitate.
Phloroglucin and Hydrochloric Acid.	not colored	cherry-red.	....	....	....	....	....
Hanstein's Aniline-violet (equal parts of Methylviolet and Fuchsine).	not at all, or very slightly colored.	violet	dark violet.	....	blue-violet.	fox-red	....

INDEX.

Calyculus, 70
Calyptrogen, 175
Calyx, 68
  leaves, 67, 68
Cambiform, 217
  cells, 231
Cambium, 220
  activity of, 221
  intrafascicular, 222
  ring, 220
Campylotropous, 90
Canals, schizogenic, 254
Cane sugar, 141
Caoutchouc, 249
Capitulare, 24
Capitulum, 79
Capsule, 85
  wall of, 157
Carbohydrates, 235
Carmine solution, 278
Carpel, 71, 73, 76, 84
Carpophore, 83
Caruncle, 90
Caryopsis, 84
Catalonians, 29
Cataphylla, 66
Catkin, 78
Cato, 21
Caulis, 62
Caustic soda, 271
Cavities, 235
  mucilage, 244
Cecidiæ, 265
Cell, 93
  aggregates, 174
  change of form of, 148, 171
  contents of, 94
  fusions, 173, 217
  growth of, 148
  membrane, 148
  membrane, chemical behavior
    of, 160
  nucleus, 96
  passages, 241
  sap, 94
  wall, 94, 148
Cells, bast, 156, 171, 194
  collenchyma, 194
  cork, 187
  crystal, 223
  daughter, 94
  epidermal, 181
  guardian, 239
  isodiametric, 149

Cells, latticed, 217
  lines of, 244, 249
  mucilage, 244
  multiplication of, 94
  of leaf, 172
  organized contents of, 139
  resin, 263
  secerning, 254
  secreting, 244
  spherically polyhedral, 149
  stone, 156, 200
  ·wood, 218, 223
  wood-parenchyma, 218, 223
Cellular tissue, 174
Cellulose, fungus, 160
  membrane, 159
  starch, 121
Central rhizome, 59
Centric leaves, 210
Cerasin, 164
Cerin, 161
  reaction, 269
Chalaza, 89
Changes of the drug on drying,
  140
Charaka, 20
Charlemagne, 24
Chemical constituents, 16
Chinese, 19
Chloride of iron, 275
  of zinc with iodine, 275
Chlorine, 145
Chloroform, 274
Chlorophyll, 100
  bodies, 100
  coloring matter, 100
  crude, 100
  formation of, 103
  granules, 100
  granules, fundamental mass
    of, 100
  pure, 100
  reactions, 104
  spectrum, 101
Chlorophyllan, 101
Chlorosis, 144
Choripetalous, 68
Choriphyllous, 68
Chorisepalous, 68
Chorisis, 77
Chromic acid, 268
Cincinnus, 64, 79
Circinal vernation, 69
Clusius, 31

CORRIGENDA.

Page 170, for *lacticiferous*, read *laticiferous*.
Page 172, notes 2 and 5. Here and in a few other places the Greek accents have been broken off during the printing.
Page 211, second line from top, for *falling off* of the membrane, read *folding of* the membrane.
Page 235, sixth line from top, for *tubes*, read *tubers*.

www.ingramcontent.com/pod-product-compliance
Lightning Source LLC
Chambersburg PA
CBHW021506210326
41599CB00012B/1145